Computational Fluid Dynamics in Practice

Edited by

N Rhodes

Professional
Engineering
Publishing

Published by Professional Engineering Publishing Limited,
Bury St Edmunds and London, UK.

First Published 2001

ISBN 978-1-86058-352-0

A CIP catalogue record for this book is available from the British Library.

Contents

Authors' Details

Details of contributing authors are listed below.

Chapter 1 – The Issue of Numerical Accuracy in Computational Fluid Dynamics
D Drikakis, Department of Engineering, Queen Mary, University of London, UK

Chapter 2 – Detection of Multiple Solutions using a Mid-cell Back Substitution Technique Applied to Computational Fluid Dynamics
S R Kendall and *H V Rao,* Department of Engineering, The University of Huddersfield, UK

Chapter 3 – A Comparison of a Conventional RANS and a Lattice Gas Dynamics Simulation – a Case Study in High-speed Rail Aerodynamics
A Gaylard, Aerodynamics and CFD, MIRA, Nuneaton, UK

Chapter 4 – Mesh Generation – the Ricardo Philosophy
S M Sapsford and *M E A Bardsley,* Ricardo Consulting Engineers Limited, Shoreham-by-Sea, UK

Chapter 5 – The Validation of Rapid CFD Modelling for Turbomachinery
H-H Tsuei, K Oliphant, and *D Japikse,* Concepts NREC Inc., Sandy, Utah, USA

Chapter 6 – Numerical Determination of Windage Losses on High-speed Rotating Discs
S Romero-Hernández, M R E Etemad, and *K R Pullen,* CASE Section, Department of Mechanical Engineering, Imperial College of Science, Technology and Medicine, London, UK

Chapter 7 – Computational Fluid Dynamics for Gas Turbine Combustion Systems – where are we now and where are we going?
K Menzies, Combustion Systems – Engineering, Rolls-Royce plc, Bristol, UK

Chapter 8 – Using CFD to Investigate Combustion in a Cement Manufacturing Process
D Giddings, S J Pickering, K Simmons, and *C N Eastwick,* School of Mechanical, Materials, Manufacturing Engineering, and Management, The University of Nottingham, UK

Chapter 9 – Validation of the Coal Combustion Capability in the Star-CD Code
A Ghobadian and *F Lee,* Computational Dynamics Limited, London, UK; *P Stephenson,* Innogy plc, Swindon, UK

Chapter 10 – Blast Wave Simulation
M Docton, S Rees, and *S Harrison,* Frazer-Nash Consultancy Limited, Dorking, UK

Chapter 11 – Built Environment Simulations using CFD
D Woolf and *G Davies,* Ove Arup and Partners, London, UK

Chapter 12 – Using CFD in the Design of Electric Motors and Generators
S J Pickering, D Lampard, J Mugglestone, and *M Shanel,* School of Mechanical, Materials, Manufacturing Engineering, and Management, The University of Nottingham, UK; *D Birse,* ALSTOM Electrical Machines Limited, Rugby, UK

Chapter 13 – CFD Modelling of a Two-phase Mixing/Separation Flow
W M Dempster, Department of Mechanical Engineering, University of Strathclyde, Glasgow, UK

Chapter 14 – CFD Computation of Air–Oil Separation in an Engine Breather
I Care, Rolls-Royce plc, Derby, UK; *C N Eastwick, S Hibberd, K Simmons,* and *Y Wang,* University of Nottingham Rolls-Royce UTC, UK

Chapter 15 – Cavitation in a Pressure-activated Ball Valve
F G Mendonca, Computational Dynamics Limited, London, UK

About the Editor

Dr Norman Rhodes, BSc MSc PhD FIMechE CEng, is Divisional Director at Mott MacDonald, and head of Mott MacDonald's Simulation and Modelling Group. He is responsible for the application of flow simulation techniques to solve engineering design problems. He has been actively involved in CFD work for over 20 years. Major areas of work include fire and smoke control, complex ventilation system evaluation, and fluid dynamics of heat-exchange processes.

Dr Rhodes is Chairman of the Institution of Mechanical Engineers 'Energy Transfer and Thermofluid Mechanics Group' and is Secretary of the Permanent International Association of Road Congresses (PIARC) working group on 'Fire and Smoke Control in Tunnels'.

Related Titles of Interest

Title	Editor/Author	ISBN
Power Transmission and Motion Control (PTMC 2000)	C R Burrows and K A Edge	1 86058 264 8
Power Transmission and Motion Control (PTMC '99)	C R Burrows and K A Edge	1 86058 205 2
Introductory Guide to Industrial Flow	R Baker	0 85298 983 0
Introductory Guide to Flow Measurement	R Baker	0 85298 670 X
Improving Maintainability and Reliability through Design	G Thompson	1 86058 135 8
Hydrocyclones '96	D Claxton, L Svarovsky, and M Thew	1 86058 032 7
CFD in Fluid Machinery Design	IMechE Seminar	1 86058 165 X

For the full range of titles published by Professional Engineering Publishing contact:

Sales Department
Professional Engineering Publishing Limited
Northgate Avenue
Bury St Edmunds
Suffolk
IP32 6BW
UK

Tel: +44 (0)1284 724384
Fax: +44 (0)1284 718692
Website: www.pepublishing.com

Foreword

Most fluid dynamic situations are characterized by three-dimensionality, turbulence, and the interactive effects of physical processes such as combustion, heat transfer, and buoyancy. These factors are, in turn, influenced by the geometrical nature of the problem and the engineering systems which may be in place or under design. Computational Fluid Dynamics (CFD) techniques are increasingly used to study and understand such situations.

The advantage of the CFD approach is that the complex physical interactions which occur in a problem can be modelled simultaneously, and hence, their relative influence on the total behaviour understood. However, complete fundamental knowledge of all the underlying physics may not exist, and there are, therefore, inherent assumptions in the mathematical process which give rise to possible inaccuracy. With care, however, these approximations can be minimized to a level where the accuracy of CFD techniques is perfectly satisfactory for design purposes.

The availability of general-purpose CFD codes now make the application of these techniques relatively easy, but they do not supply the engineering knowledge or wisdom to apply them correctly. It is important, therefore, to understand how CFD is applied, the approximations used, and the factors which influence the accuracy of any simulation.

The purpose of the seminar held at the Institution of Mechanical Engineers on which this book is based, was to provide a forum for the discussion of challenging applications of CFD and to identify the developments in techniques which are likely to occur in the next generation of codes. The examples provided in this volume give a wide range of CFD applications, show the versatility of the techniques and provide the reader with some hints on best practice.

Validation of CFD models is greatly encouraged. It is worth reflecting that there are said to be two kinds of fluid dynamicists, those who engage themselves in numerical analysis of fluids and those who determine fluid behaviour from experiments. It is generally held that no one believes the results of a numerical analysis of fluid flow except the numerical analyst and everyone believes the results of the experiment except the experimentalist!

There is, sadly, some truth in this anecdote. Experiments are real, a numerical analysis is a simulation; experiments are difficult to reproduce and ensure correct measurement, simulations can be repeated in a more controlled way; much of engineering science is supported by experiment and observation, simulation with computers is relatively new and requires a different background knowledge, which this book goes a little way to provide.

Dr Norman Rhodes
c/o HNTB, USA
June 2001

1

The Issue of Numerical Accuracy in Computational Fluid Dynamics

D Drikakis

Synopsis

This chapter reviews various issues regarding the accuracy and efficiency of numerical methods used in the simulation of incompressible and compressible flows. The methods include high-resolution and high-order discretization schemes, explicit and implicit solvers, and non-linear multigrid methods. The potential to apply these methods as an 'implicit modelling' approach in the simulation of transitional and turbulent flows of engineering interest, is also discussed in contrast to the current turbulence modelling practices.

1.1 Introduction

The accuracy and efficiency of computations of transport phenomena depend strongly on the numerical properties of the discretization scheme and iterative solver employed for solving the system of equations describing these phenomena. The practice during code development is to check the accuracy of the methods/models by comparing the numerical results with analytic solutions, wherever possible, and/or experimental data, if available. Analytic solutions exist only for a very limited number of simple flows. Experiments specifically designed for Computational Fluid Dynamics (CFD) validation purposes are expensive and time-consuming to be set up. Yet, they are not free of shortcomings and difficulties due to instrumentation constraints.

To increase the reliability of numerical simulations the common practice is to obtain grid and time-step independent solutions. This can possibly be achieved for two-dimensional laminar flows, but it would be a hopeless task for three-dimensional, transitional, and turbulent flows. The difficulties to obtain reliable solutions are even more severe in the case of time-dependent simulations because the grid-independence investigations should always be accompanied by time-step independence studies. The above difficulties impose stringent constraints in our ability to provide reliable results in short turnaround times, the latter being extremely important in an industrial design environment, but they also motivate research in the two directions:

(i) To develop high-resolution and high-order methods *in both space and time*, which would enable us to attain the desired accuracy without resorting to very fine meshes that are not affordable with current computers;

(ii) to develop advanced numerical solvers such as non-linear multigrid methods, for accelerating the solution of the fluid flow equations.

There are indications (1)–(3) emerging from research on numerical methods carried out over the past two decades, but more intensively over the past few years, which conspire to the fact that a class of fluid dynamics numerical algorithms called non-linear monotone, high-resolution, and high-order methods can provide accurate results in coarsely resolved simulations. Most of the non-linear monotone schemes originate from Godunov's method (4) and were originally designed to capture shocks and other discontinuities (5)–(17) while more recently, Godunov-type schemes have also been developed for incompressible flows (18)–(21). The implementation of these schemes in conjunction with non-linear multigrid solvers, e.g. (22), implicit relaxation methods, e.g. (23), and/or high-order explicit solvers, e.g. (24), can lead to fast solutions of physically complex flows even on personal computers or desktop workstations.

The chapter presents an overview of the above methods and discusses by means of various computational examples, from the author's research work, the current status and challenges that lie ahead.

1.2 High-resolution and high-order of accuracy

1.2.1 Governing equations

The physics of incompressible and compressible flows is governed by the Navier–Stokes equations. These equations can be solved by considering the coupled generalized conservation laws, i.e. the continuity, momentum, and energy equations

$$\frac{\partial \rho}{\partial t}+\nabla\cdot(\rho \mathbf{v})=0 \tag{1.1}$$

$$\frac{\partial \rho \mathbf{v}}{\partial t}+\nabla\cdot(\rho \mathbf{v}\mathbf{v})=-\nabla\cdot\mathbf{P} \tag{1.2}$$

$$\frac{\partial e}{\partial t}+\nabla\cdot(e\mathbf{v})=-\nabla\cdot(\mathbf{v}\cdot\mathbf{P})-\nabla\cdot\mathbf{q} \tag{1.3}$$

where $\mathbf{v}$, ρ, e, and $\mathbf{q}$ stand for the velocity components, density, total energy per unit volume, and heat flux, respectively. The pressure tensor $\mathbf{P}$ is defined by

$$\mathbf{P}=p(\rho,T)\mathbf{I}+\frac{2}{3}\mu(\nabla\cdot\mathbf{v})\mathbf{I}-\mu\left[(\nabla\mathbf{v})+(\nabla\mathbf{v})^{T}\right] \tag{1.4}$$

where, $p(\rho,T)$ is the scalar pressure, $\mathbf{I}$ is a unit diagonal tensor, T is the temperature, and μ is the dynamic viscosity coefficient. To complete the above system, an equation of state (EOS) for the scalar pressure must be employed. In the case of a perfect gas the EOS is: $p=\rho RT$, where R is the gas constant.

The coupled system of the above equations can also be written in a matrix form, for a general curvilinear co-ordinates system as

$$(JU)_t + (E_I - E_V)_\xi + (F_I - F_V)_\eta + (G_I - G_V)_\zeta = \mathcal{H} \tag{1.5}$$

where the indices I and V stand for the inviscid and viscous fluxes, respectively. $U = (\rho, \rho\,\mathbf{v}, e)^T$ is the unknown solution vector; E, F, and G are the fluxes that contain the inviscid and viscous terms; ξ, η, and ζ are the curvilinear co-ordinates; J is the Jacobian of the grid transformation from Cartesian to curvilinear co-ordinates, and $\mathcal{H}$ contains source terms arising from external forces, turbulence modelling, chemical reactions etc. When no physical diffusion effects are considered, such as viscosity, thermal conduction or molecular diffusion, the fluxes with the subscript V can be dropped thus obtaining the Euler equations.

1.2.2 Monotonicity, conservation, and bounded total variation

The numerical solution of the Euler and Navier–Stokes equations requires careful discretization of the spatial and time derivatives. It is broadly accepted that the numerical accuracy is mainly affected by the discretization of the advective terms (inviscid fluxes in (**5**)). There are a number of different methods that can be utilized to discretize these terms but regardless of their details, the accuracy is dependent on four fundamental numerical properties:

(i) Monotonicity;
(ii) Total Variation Diminishing (TVD);
(iii) Conservation; and
(iv) Entropy condition.

The latter is primarily associated with the computation of flows containing discontinuities. A numerical scheme is considered as monotone if it does not lead to an oscillatory behaviour of the numerical solution. The monotonicity condition is also strongly linked to the concept of bounded total variation which says that the total variation, (TV)

$$TV = \int \left| \frac{\partial u}{\partial x} \right| dx$$

of any physically admissible solution u does not increase in time. Conservation implies that the mass, momentum, and kinetic energy integrated over the computational space do not change due to the algorithm. Finally, the *entropy condition* is related to the fact that certain numerical schemes provide solutions not physically acceptable, since they are associated with a decrease in entropy which is not, however, allowed by the second principle of thermodynamics.

The above properties are strongly linked to two numerical effects that can be introduced in the numerical solution, namely the *numerical dispersion* and *numerical diffusion* (dissipation). From the computational view point, the challenge is to design numerical methods which would minimize the above effects. Spectral methods (**25**) have been used in flows with simple geometries in order to reduce numerical dissipation. However, apart from the difficulty of extending these methods for complex geometries, there exist additional difficulties associated with aliasing errors resulting from the evaluation of non-linear terms on a discrete grid as well as from the Gibbs phenomenon of oscillations, especially in the case of flows with discontinuities. These effects become even more important under the presence of steep gradients and have been extensively discussed in the literature (**25**), (**26**).

A promising route for developing accurate numerical methods which can also be applicable to engineering flows, is the non-linear monotone schemes in the context of finite volume methods. The state of the art of these schemes aiming at satisfying the aforementioned properties are also referred to as *high-resolution* and *high-order* methods. These methods emerge from the theory of hyperbolic conservation laws and the original Godunov method (**4**). The latter is the forerunner of all the well known approximate Riemann solvers, e.g. (**7**), (**8**), upwind Godunov schemes, e.g. (**6**), (**10**), (**11**), (**14**), TVD methods, e.g. (**9**), (**12**), (**13**), (**15**), and of schemes with order of accuracy higher than two (see discussion in Section. 1.2.4).

1.2.3 High resolution

High-resolution schemes are those with the following properties (**9**), (**17**):

(i) Provide at least second order of accuracy in smooth areas of the flow;
(ii) Produce numerical solutions free from spurious oscillations; and
(iii) In the case of discontinuities, the number of grid points in the transition zone containing the shock wave is smaller in comparison with that of first-order monotone methods.

High-resolution can be realized by numerically reconstructing the variables at the cell faces of a computational volume via the solution of a Riemann problem. This can be achieved by using Godunov-type methods which offer an ingenious computational framework for developing schemes that adjust the amount of numerical dissipation locally, i.e. at the cell faces, in order to maintain monotonicity and conservation.

The reconstruction of the variables should be at least second-order accurate and can be obtained by deriving average procedures for the calculation of the unknown variables, which take into account the characteristic signal velocities. The latter are defined by the eigenvalues of the Euler equations. In the case of multi-dimensional and multi-physics problems the reconstruction can be obtained not only for the velocity, energy, as well as density and/or pressure variables, but also for the unknown variables of any additional partial differential equations arising from modelling considerations and conservation laws describing the multi-physics nature of a problem.

High-resolution Godunov-type methods can provide accurate results in complex unsteady flows even when coarse grids are employed. In Fig. 1.1 we present results from simulations (**16**) of unsteady shock-wave diffraction over a cylinder. A series of investigations regarding shock-wave diffraction have been performed by the author and his collaborators over the past few years (**16**), (**27**), (**28**), due to the importance of this problem in explosion dynamics. The computations in (**16**) were performed on coarse grids without local grid-refinement, using hybrid upwind Godunov-type methods. The predictions for the pressure loads on the cylinder surface were compared with the corresponding results from unstructured adaptive-grid computations (**27**). As can be seen from Fig. 1.1, without using grid adaptation the results obtained by the Godunov-type methods on coarse grids are comparable to those obtained by utilizing adaptive grids.

Two further examples from the author's work on high-resolution Godunov-type methods for compressible and incompressible flows are given below. In (**23**), the generalized Riemann problem has been considered for the case of compressible turbulent flows; the latter being modelled by using statistical turbulence models in the context of two and three-equation linear and non-linear eddy-viscosity models, e.g. k–ε and k–ω models, where, k, ε, and ω stand for the turbulent kinetic energy, turbulent dissipation rate, and specific turbulent

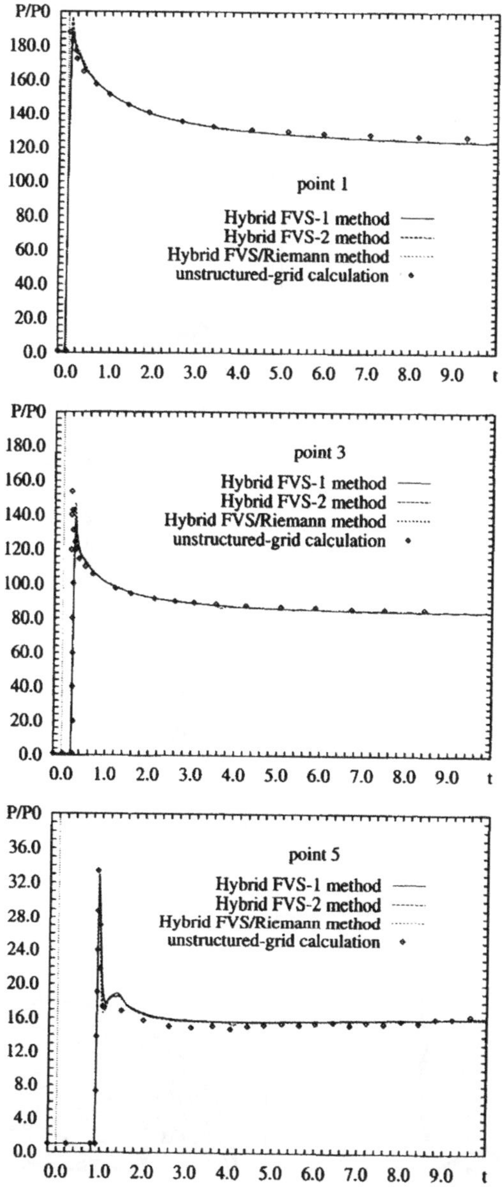

Fig. 1.1 Unsteady shock-wave diffraction around a cylinder at an incident Mach number M_{in} =5;

(a) wall pressure histories at different positions around the cylinder using:

(i) different Godunov-type methods (16) and a coarse structured-based grid without adaptation and

(ii) unstructured adaptive grids (27).

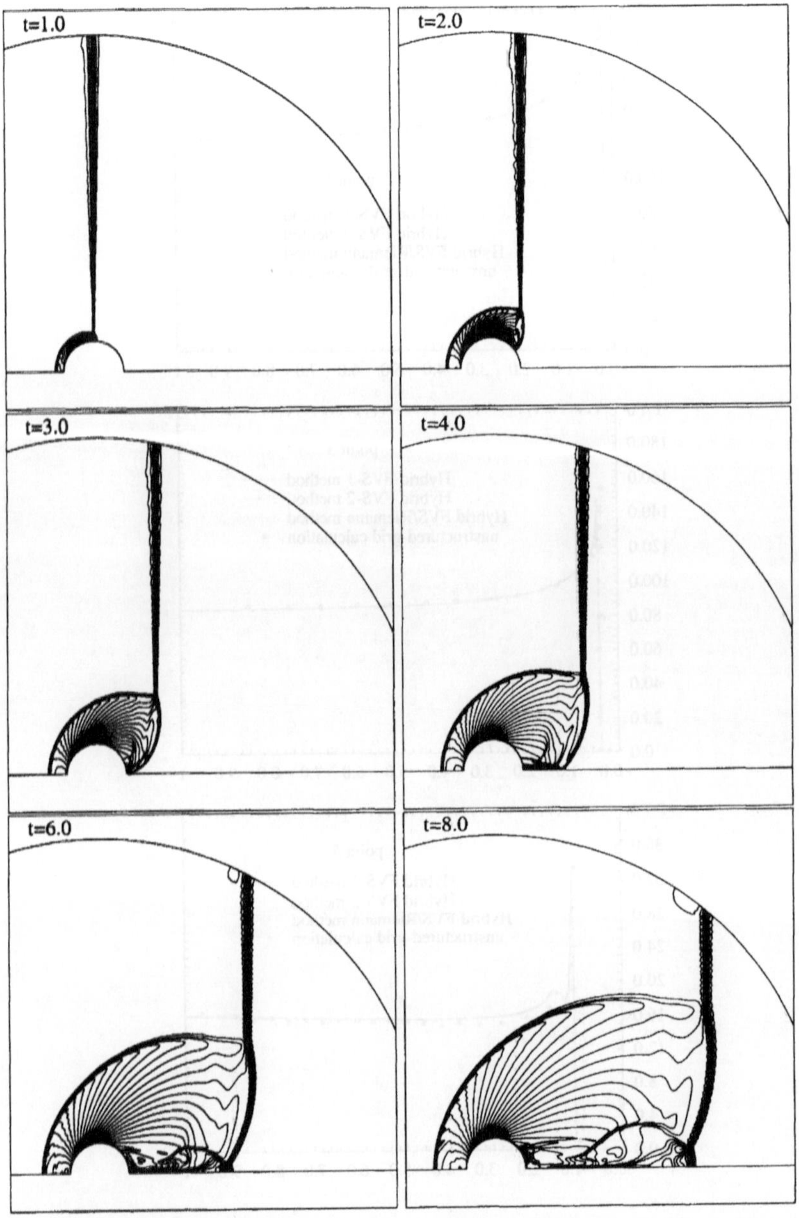

Fig. 1.1 Unsteady shock-wave diffraction around a cylinder at an incident Mach number M_{in} =5;
(b) isodensity contours at different time instants

dissipation rate, respectively. According to (**23**), the reconstructed variables (denoted below by tilde), for a two-dimensional Cartesian co-ordinate system[1] by

$$\begin{pmatrix} \tilde{\rho} \\ (\widetilde{\rho u}) \\ (\widetilde{\rho v}) \\ \tilde{e} \\ (\widetilde{\rho k}) \\ (\widetilde{\rho \epsilon}) \end{pmatrix} = \begin{pmatrix} \rho_0 + R_1 + R_2 \\ (\rho u)_0 + (u+s)R_1 + (u-s)R_2 \\ (\rho v)_0 + v(R_1 + R_2) \\ e_0 + (H + s\lambda_0)R_1 + (H - s\lambda_0)R_2 \\ (\rho k)_0 - kR_1 - kR_2 \\ (\rho\epsilon)_0 - \epsilon R_1 - \epsilon R_2 \end{pmatrix} \tag{1.6}$$

where H is the total enthalpy of the fluid and λ_0 is the zeroth eigenvalue. The terms R_1 and R_2 are functions of the mean, ρ_j, $(\rho u)_j$, $(\rho v)_j$, e_j, as well as of the turbulent, $(\rho k)_j$, $(\rho e)_j$, flow variables defined on the three characteristics denoted by the index $j = 0, 1, 2$. The functions R_1 and R_2, as well as the reconstructed variables in equation (1.6), are analytically derived (**23**). The advective fluxes E_I, F_I, and G_I in equation (1.5) are calculated using the tilde variables (1.6). Reconstructions such as the above provide better accuracy as well as direct coupling of the Navier–Stokes and turbulence transport equations through the functions R_1 and R_2. Results for both steady and unsteady flows using the above Godunov-type scheme can be found in (**23**), (**29**)–(**30**).

Similar Godunov-type reconstructions can also be constructed for incompressible flows. The incompressible Navier–Stokes equations are derived under the assumption of constant density flow. In this case, the continuity and momentum equations are decoupled because no pressure or density terms appear in the former. One way to directly couple the equations is to use the artificial-compressibility (AC) approach proposed by Chorin (**31**). A pseudo-time derivative of pressure, $\beta^{-1}\partial p/\partial\tau$ (β is the artificial compressibility parameter), is added to the continuity equation thus providing coupling between continuity and momentum equations in the case of constant density flows. For steady flows the pseudo-time pressure derivative will become zero as the steady state is approached. The AC approach can also be extended to time-dependent flows via dual-time stepping (**21**), (**32**)–(**34**).

Various Godunov-type methods for incompressible flows have been presented in the literature (**18**), (**19**), (**21**). The characteristic-based method of (**19**), (**21**) defines the primitive variables for a two-dimensional Cartesian co-ordinate system[2] by

$$\begin{pmatrix} \tilde{p} \\ \tilde{u} \\ \tilde{v} \end{pmatrix} = \frac{1}{2s}\begin{pmatrix} \lambda_1\lambda_2(u_2 - u_1) + \lambda_1 p_2 - \lambda_2 p_1 \\ p_1 - p_2 + \lambda_1 u_1 - \lambda_2 u_2 \\ 2sv_0 \end{pmatrix} \tag{1.7}$$

where $\lambda_0 = u$, $\lambda_1 = \lambda_0 + s$, $\lambda_2 = \lambda_0 - s$, and $s = \sqrt{\lambda_0^2 + \beta}$. Similarly to equation (1.6), in equation (1.7) the indices also denote the corresponding characteristics. The variables on the

1 The definition of the reconstructed compressible flow variables for a curvilinear co-ordinate system can be found in (**23**).

2 The definition of the reconstructed incompressible flow variables for a curvilinear co-ordinates system can be found in (**19**) and (**21**) for two-dimensional and three-dimensional problems, respectively.

characteristics are defined by second-, or higher-order interpolation from the values in the neighbouring cells (19), (21). Similarly to the compressible flow case, the advective fluxes of the incompressible equations are calculated using the tilde variables of equation (1.7).

Results from the implementation of the above method in incompressible flows featuring instabilities and transition to turbulence (35), (36) are discussed below. The study of two- and three-dimensional instabilities in channel flows at relatively low Reynolds numbers is motivated by the increasing interest in understanding transport phenomena appearing in nano-engineering applications. In Fig. 1.2, we present the development of instabilities at different Reynolds numbers for two-dimensional suddenly-expanded flows (35); the instabilities appear as asymmetric separation, though the geometry as well as the initial and boundary conditions are perfectly symmetric. Our results are in agreement with the stability analysis of this flow (37). We also found (35) that the onset of the instability is affected by the order of interpolation employed in calculating the characteristic variables (35). The best results were achieved for a third-order Godunov-type interpolation.

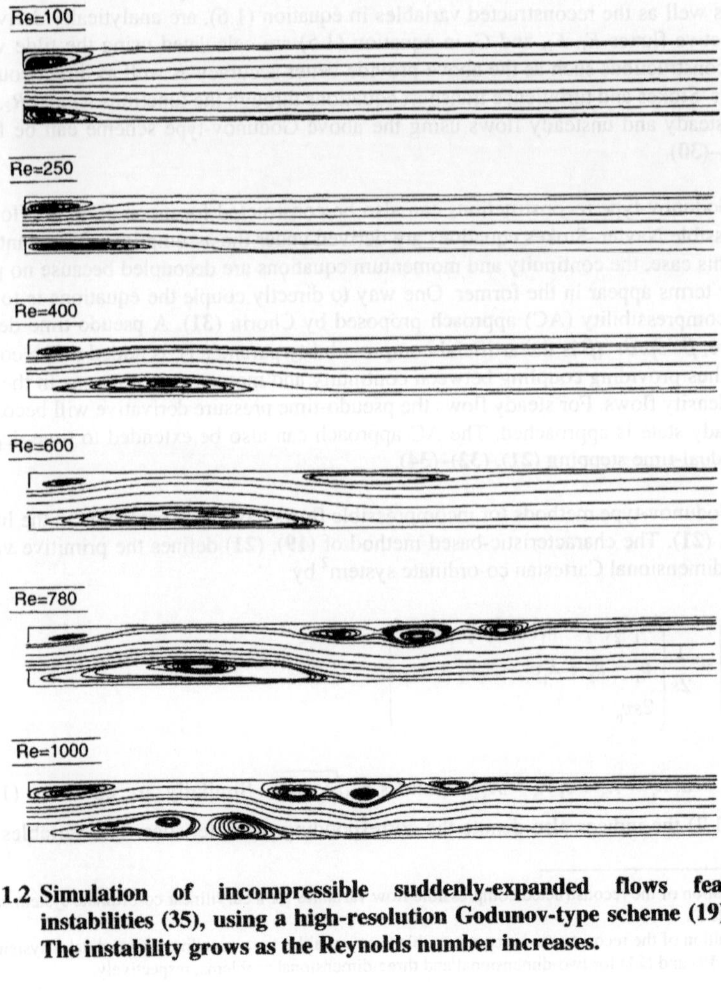

Fig. 1.2 Simulation of incompressible suddenly-expanded flows featuring instabilities (35), using a high-resolution Godunov-type scheme (19), (20). The instability grows as the Reynolds number increases.

More recently, we have also conducted investigations of three-dimensional instabilities and laminar-to-turbulent transition in pulsatile flows through a stenosis (**36**). In the context of biofluid mechanics, the computational study of flows through stenoses is motivated by the need to obtain a better understanding of the impact of flow phenomena on diseases such as atherosclerosis and stroke. These phenomena include asymmetric flow separation, instabilities, laminar-to-turbulent transition and turbulence; the above have profound effects on the wall-shear stress distribution. Yet, understanding of these phenomena can potentially lead to the development of better diagnostic criteria and clinical detection techniques, e.g. ultrasound. In Fig. 1.3, we show the stenosis model used in our investigation (**36**) as well as the development of the instability downstream of the stenosis.

In this section the discussion focused on high-resolution Godunov-type methods that provide second-order of accuracy[3]. In the next section, we discuss various ways to increase the global accuracy of the discretization beyond second-order.

1.2.4 High-order of accuracy

High-order methods aim to increase the accuracy and, simultaneously, avoid spurious oscillations. The investigation of numerical schemes with order of accuracy higher than two was initiated in the 1980s by Harten *et al.* (**39**) with the development of the essentially non-oscillatory schemes (ENO). Most of the research in this area has been performed in connection with the one-dimensional equations of gas dynamics (**24**), (**40**).

The basic idea behind ENO schemes is to avoid growth of spurious oscillations by defining the fluxes within an interpolating stencil in such a way that the solution is, possibly, the smoothest one. Using the ENO approach the accuracy of the flux calculation can go beyond second-order of accuracy but the implementation of these schemes is not trivial at all.

Here, we briefly discuss the uniformly high-order (UHO) scheme of (**21**). This scheme also draws from the ideas underlying the essentially nonoscillatory (ENO) approach (**24**), (**39**). It aims at increasing the accuracy of the intercell fluxes via a high-order interpolation (flux reconstruction) procedure that can be briefly summarized as follows.

The characteristic-based scheme of (**19**) is initially used to provide a first approximation for the advective fluxes, e.g. $(E_I)_i$, at the cell centers i. Then, the cell-centered approximated fluxes are interpolated to provide high-order accurate left (E_L) and right (E_R) intercell fluxes. For example, the E_R flux is defined by

$$(E_R)_{i+1/2} = \sum_{k=-r+3-n}^{r-2} \alpha_k^r E_{i+k} \tag{1.8}$$

where r denotes the order of accuracy of the resulting scheme, with $n = 0\ \forall\ r > 3$ and $n = 1$ if $r = 3$; and the coefficients α_k^r are *constant* weights, defined by an analytic procedure that minimizes the numerical dissipation and dispersion (**21**). This high-order interpolation can be retained throughout the computations only in the case of periodic boundaries; in the vicinity of solid boundaries the second-order-accurate scheme would be used. Finally, the left and right intercell fluxes can be combined by using the Lax-Friedrichs (**41**), Roe (**7**), or (**42**)

3 We refer to second-order of accuracy in both space and time. We also note that some methods, e.g. (**15**), provide directly second-order of accuracy in space and time, where in other cases, e.g. the characteristic-based scheme (**19**), (**21**), the global accuracy is achieved only by combining the Godunov-type method with high-order TVD Runge–Kutta schemes (**24**).

schemes to calculate the new intercell flux $E_{i+1/2}$ which is subsequently used in the discretization of the advective flux derivative, i.e. $\partial E / \partial x = (E_{i+1/2,j} - E_{i-1/2,j}) / \Delta x$.

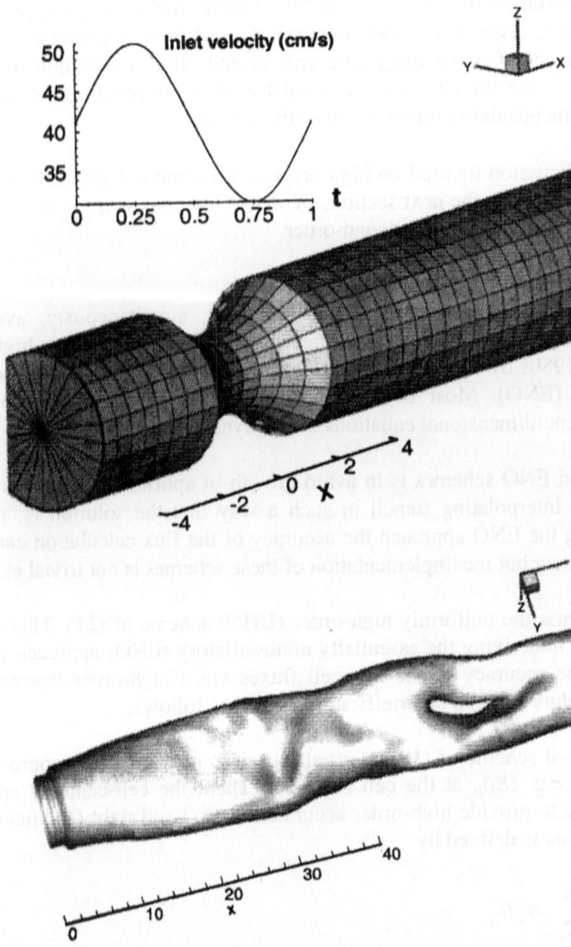

Fig. 1.3 Study of incompressible three-dimensional pulsatile flow through a stenosis (36) using a high-resolution Godunov-type scheme (19), (20). The upper plot shows the stenosis model and inlet waveform while the lower plot shows, by means of the velocity isosurface *u* = 0.92, the development of the instability downstream of the stenosis. The instability results in complete breaking of the axisymmetry of the flow and transition to turbulence.

The third-order version of the method has been recently implemented with success in incompressible flows featuring complex vortical structures (3), (21), and some results are presented below. In Fig. 1.4, the instantaneous solutions from the simulation of a periodic double mixing layer at *Re*=5000 are compared using three different numerical methods:

(i) The uniformly high-order scheme (UHO) (**21**) in conjunction with a characteristic-based method (CBM) (**19**), (**20**);

(ii) The first-order; and

(iii) Second-order Rusanov flux (**42**).

The second-order flux is utilized in conjunction with the MUSCL scheme (**6**).

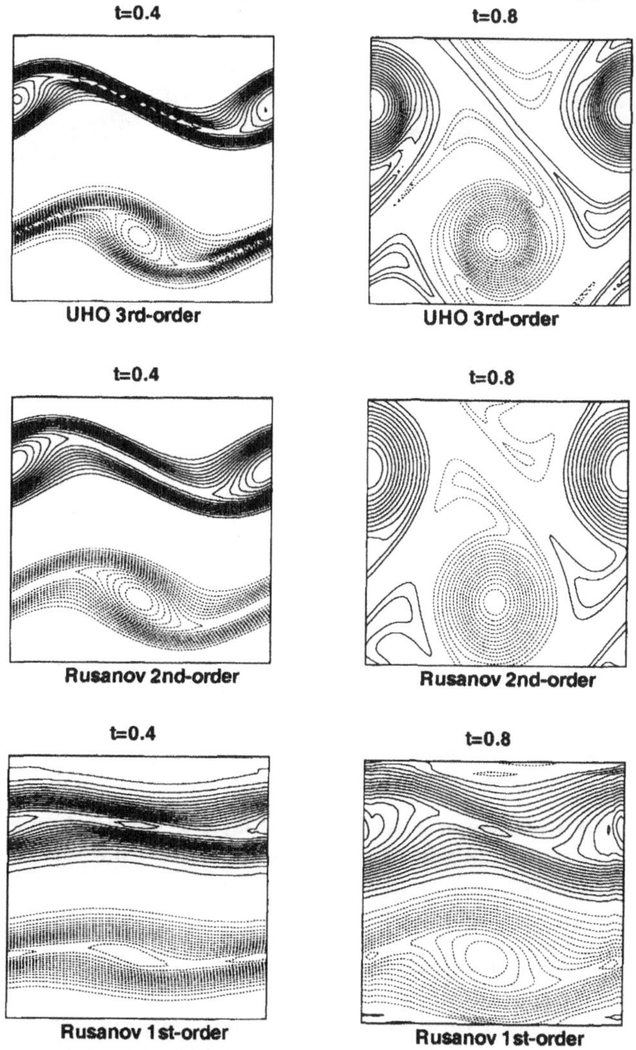

Fig. 1.4 Instantaneous plots from simulations (21) of a double mixing layer (Re = 5000; 64 × 64 grid). Comparison of the results obtained by the 3rd-order uniformly high-order (UHO) – characteristic-based method (CBM) (21) with the corresponding results obtained by first- and second-order accurate Godunov-type schemes (42).

The results of Fig. 1.4 have been obtained on a (coarse) 64×64 grid in order to examine the accuracy of the schemes in under-resolved simulations. The flow evolution on a 128×128 grid is shown in Fig. 1.5. The results reveal that the increased dissipation exhibited by first- and second-order schemes leads to a misrepresentation of the flow structures in the case of under-resolved simulations, while using higher-order discretization most of the flow features are captured even on the coarse grid.

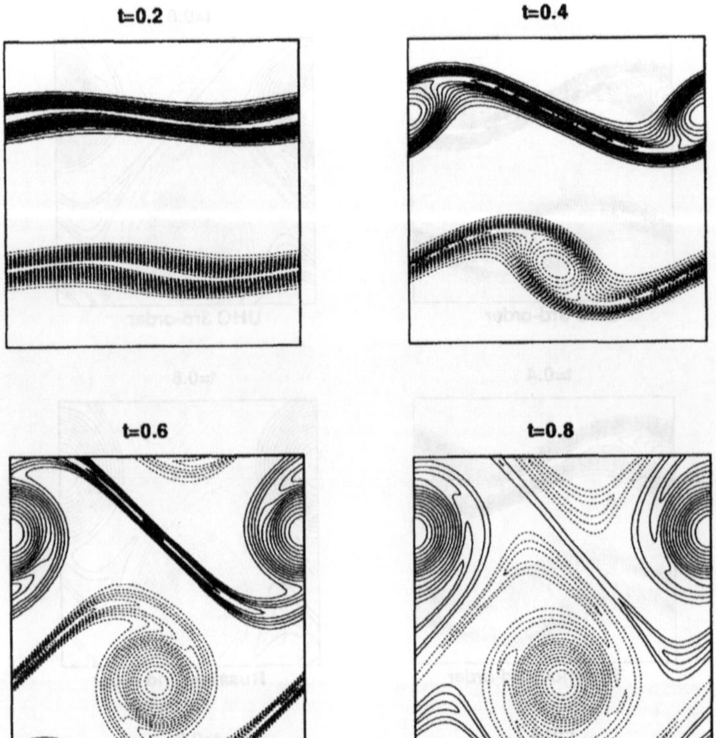

Fig. 1.5 Evolution of a double mixing layer as predicted on a 128×128 grid by the third-order UHO-CBM scheme (21).

1.3. Robustness and efficiency

In addition to the accuracy issue, another central issue regarding the implementation of numerical methods is to obtain efficient solutions regardless of the complexity of the flow problem and with the minimum possible adjustments of numerical parameters, e.g. time step, under-relaxation coefficients, flux limiters, multigrid sub-iterations.

A factor that can significantly affect the performance of a CFD code is the iterative method used to march the numerical solution towards convergence, in both steady and time-dependent flows. The numerical solvers can be broadly classified into explicit and implicit methods. The choice of the method is strongly related to the physical scales needed to be resolved as well as to the time step limitations imposed by the small grid spacing in certain regions of the computational domain, e.g. near solid boundaries. A possible approach is to solve all transport equations using a Runge–Kutta explicit method. Such a scheme is relatively easy to implement but it results in CFL numbers less than 1 in theory, and about 0.5, in practice. Although explicit solvers exhibit slow convergence rates, their ease of implementation makes them very attractive, especially when they are combined with advanced convergence acceleration techniques such as multigrid methods, e.g. **(22)**, **(43)**.

A second approach is to use implicit-unfactored relaxation schemes, e.g. **(45)**, **(46)**. These provide higher CFL number than explicit and implicit-factorization methods. Concerning additional equations arising from modelling, e.g. the turbulence transport equations, one can implement them in conjunction with the fluid flow equations following a loosely-coupled approach (henceforth labelled 'implicit-decoupled'). A block-implicit solution for the Navier–Stokes equations is first obtained followed by the solution of the turbulence transport equations. Due to the decoupling of the fluid flow and turbulent transport equations the convergence rates may significantly decrease, especially in time-dependent flows.

Finally, a third approach is to solve all equations in a strongly coupled fashion using an implicit method **(23)**, **(29)**, **(46)**; this approach is labelled 'implicit-coupled'. The implicit-coupled approach provides robust solutions but it is more complicated in the development, e.g. for three-dimensional compressible flows using a two-equation turbulence model, the Jacoby and eigenvector matrices for seven equations should be derived and numerically implemented. Comparison between explicit, implicit-decoupled, and implicit-coupled solutions is shown in Fig. 1.6 for the case of compressible flow around an axisymmetric body, featuring shock/boundary-layer interaction and separation.

One particularly attractive alternative to implicit-unfactored methods is the use of non-linear multigrid schemes, e.g. **(22)**, in conjunction with explicit TVD Runge–Kutta methods **(24)**. The latter can provide third- or fourth-order accurate discretization in time and, additionally, offer a flexible framework for extending the computer code with additional differential equations in the case of multi-physics flow problems. For the case of the Navier–Stokes equations the non-linear multigrid method of **(22)** is realized via an unsteady-type procedure. The equations are not solved exactly on the coarsest grid but some pseudotime iterations are performed on the finer grids and some on the coarsest grid. Numerical experiments showed that significant acceleration can be achieved in both steady **(22)** and time-dependent flow problems **(21)**. In time accurate computations the multigrid solution of the equations is obtained by performing multigrid cycles at each real time step (Fig. 1.7). Results from the acceleration of the numerical convergence using the above method are shown in Fig. 1.8 for the case of three-dimensional flow in a channel with a 90 degree bend (see **(22)** for more details).

More recently, the development of dynamically-adaptive multigrid methods for the Navier–Stokes equations has also been presented **(44)**. The dynamically-adaptive multigrid exploits the non-uniform convergence behaviour of the numerical solution during the iterations and solves the equations only in those cells for which convergence has not yet been achieved. As a result, the size of the computational problem is adaptively and progressively reduced, and so is the computing time.

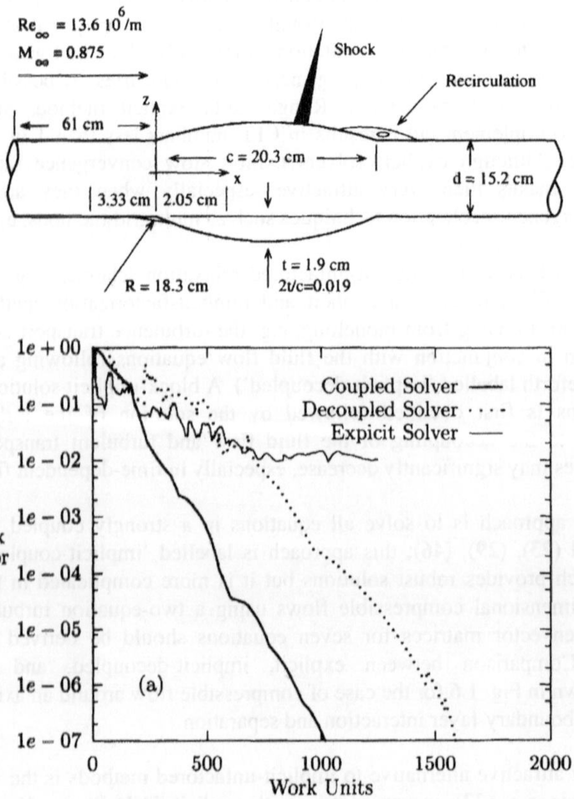

Fig. 1.6 Comparison of convergence histories between implicit-coupled, implicit-decoupled and explicit solvers for a transonic turbulent, compressible, flow featuring shock/boundary layer interaction and separation (23).

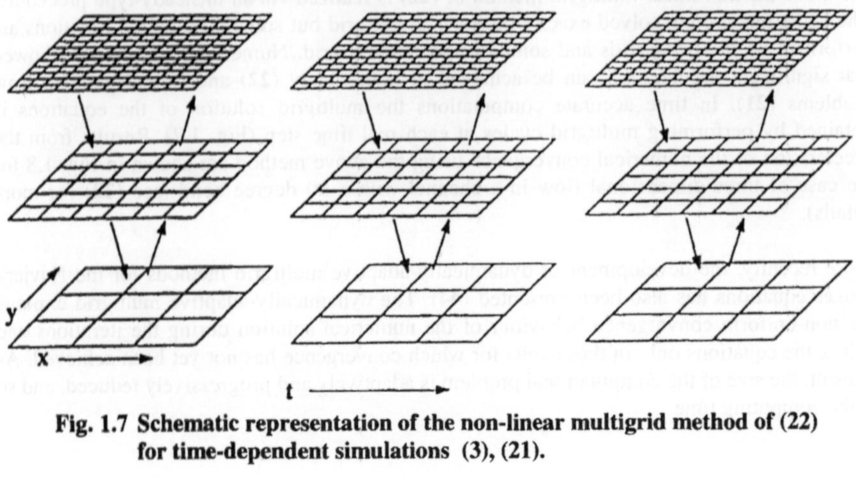

Fig. 1.7 Schematic representation of the non-linear multigrid method of (22) for time-dependent simulations (3), (21).

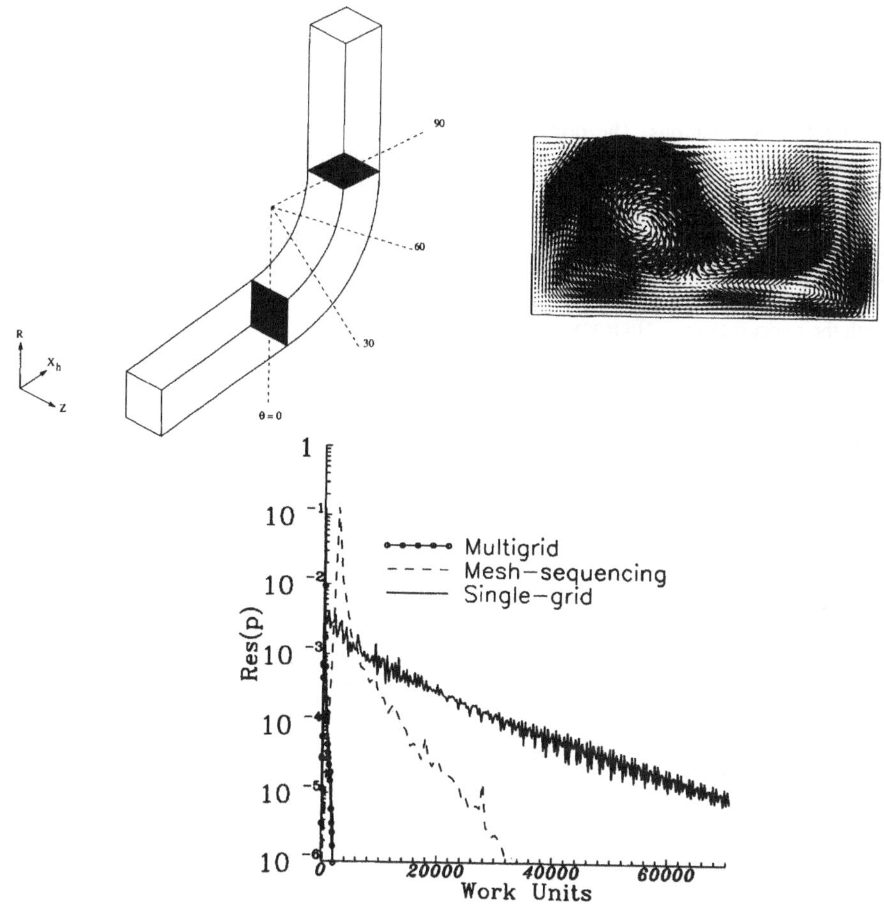

Fig. 1.8 Acceleration of the convergence using the non-linear multigrid method for three-dimensional flow in a channel with a 90 degree bend (22). The single grid and mesh-sequencing convergence histories have also been plotted for comparison purposes. The mesh-sequencing provides a better initial guess for the fine grid computations by utilizing a sequence of coarse grids (without using multigrid) and interpolating the most recent coarse-grid solution onto the fine grid (see (22) for more details).

1.4. Simulation of turbulent flows

The computation of turbulent shear flows has been mainly pursued by three different approaches:

(i) The Reynolds Averaged Navier–Stokes (RANS);
(ii) The large eddy simulation (LES);
(iii) The direct numerical simulation (DNS).

The RANS approach is the most broadly used in the context of engineering flows. So far, the main issue in connection with RANS has been the development of statistical turbulence models in the context of linear and non-linear eddy-viscosity models, and second-moment closures. In all cases additional transport equations require solution in conjunction with the Navier–Stokes equations. The primary aim of RANS turbulence modelling is to create a simpler framework for simulating flows of engineering interest. However, this is far from being the case, especially when complex models such as non-linear eddy-viscosity models (NLEVM), e.g. **(47)**, are employed. Numerical implementation of these models is not a trivial task, since the details of the implementation may have profound effects on the convergence. Complex models such as NLEVM and second-moment closures are far from being numerically robust, and significant efforts still need to be spent with respect to improving both their accuracy and efficiency.

Implementation of NLEVM in conjunction with Godunov-type methods and implicit schemes has been presented by the author and his collaborators in the recent past for both steady and unsteady compressible flows **(30)**, **(48)**. Results from dynamic-stall simulations around an oscillating aerofoil are shown in Fig. 1.9. We note that similar vortex flow phenomena occur in both oscillating and ramping aerofoils (Fig. 1.9). The results for the oscillating aerofoil correspond to subsonic flow conditions ($M_\infty = 0.2$, $Re = 10^6$) with mean incidence angle $\alpha_0 = 15$ degrees, amplitude, and reduced-frequency of oscillations $\alpha_1 = 10$ degrees and $f^* = 0.1$, respectively. The computations for the unsteady airloads have been compared with the corresponding experimental results from **(50)**. The NLEVM provides overall better results but the differences between experiments and computations are still substantial. Moreover, long computing times can be required in the case of complex models such as NLEVM or second-moment closures. It has also been found **(48)** that in many flow cases the one-equation Spalart–Allmaras model **(52)** can provide similar results to the NLEVM. This can also be seen in the results (Fig. 1.10) from transonic buffet computations **(53)**, where similar predictions for the buffet onset are obtained by the one-equation Spalart–Allmaras **(52)** and a non-linear eddy-viscosity model **(47)**.

Concerning the LES approach the main issue is to account for the unresolved small scales. In the context of LES the unresolved structures are represented by a subgrid scale model (SGS). The modelling difficulties, particularly in wall bounded flows, seem to be similar to those encountered in the RANS approach. Additionally, LES requires fully three-dimensional unsteady computations to be performed, though some successful two-dimensional simulations of flows with large separation have also been reported **(55)**. Yet, the classical LES approach resorts to very fine meshes, though coarser than those used in DNS. The objective of DNS is to resolve all scales of turbulent motion down to the Kolmogorov eddy. Adequate resolution must ensure that simulated structures are correct and not numerical artifacts. However, currently DNS for turbulent flows of engineering interest is not feasible due to the lack of adequate computing resources.

In the case of simulations of complex engineering flows the question is also whether one needs to simulate the flow down to the smallest scale. The Kolmogorov spectrum **(56)** describes how the energy density of turbulent structures decreases rapidly with increasing the wave number, where the Kolmogorov scale is the scale at which the viscous dissipation dominates the inertial flow of the fluid. The downward transfer of energy from large to small scales is called the turbulent cascade process. The latter stops at the Kolmogorov scale, where an eddy is so small that it diffuses rapidly. Previous computations, experiments and theoretical analysis (see e.g. **(1)** has shown that the physics of the turbulent cascade is controlled by the macroscopic scales of the flow and the process of dissipation of this energy due to molecular viscosity takes place primarily at scales considerably larger than the Kolmogorov scale. Another important issue is that the energy transfer is dominated by local interactions. In other words, the energy does not skip from the large to the small scales, but the energy extraction from a given scale occurs as a result of interactions with eddies no more than an order of magnitude smaller.

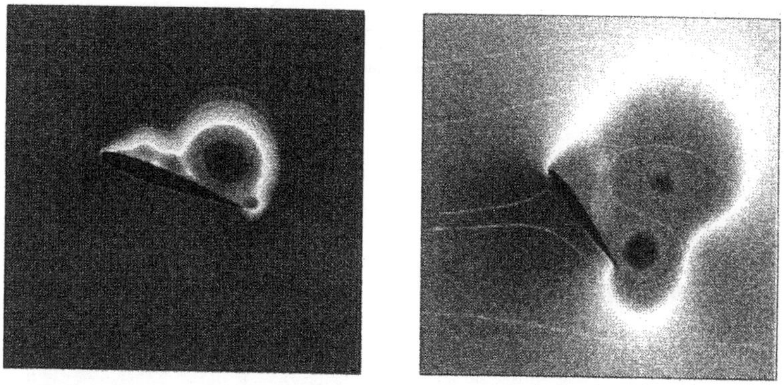

Oscillating Aerofoil

Ramping Aerofoil

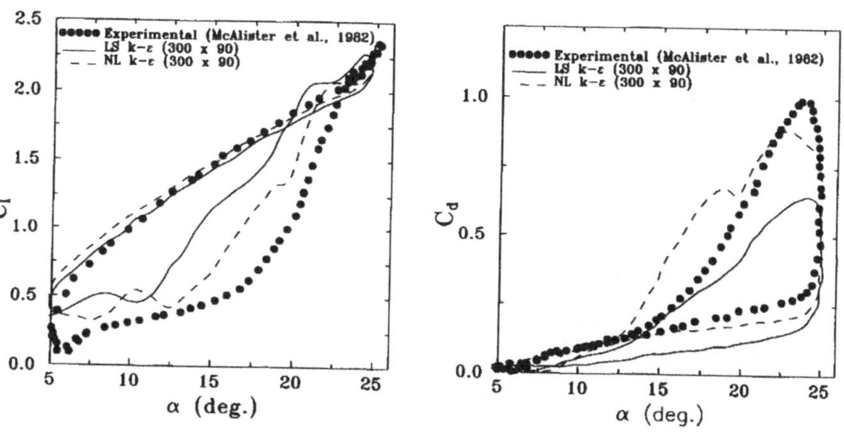

Fig. 1.9 Results from dynamic stall simulations (48), (49) around a NACA 0012 aerofoil at subsonic deep-stall flow conditions. The upper plots show the density field at maximum incidence for an oscillating and a ramping NACA-0012 aerofoil (see (48), (49) for more details). The lower plots compare the results between a linear (LS) (51) and a non-linear EVM (NL) (47) for the lift and drag coefficients. The experimental results are from (50).

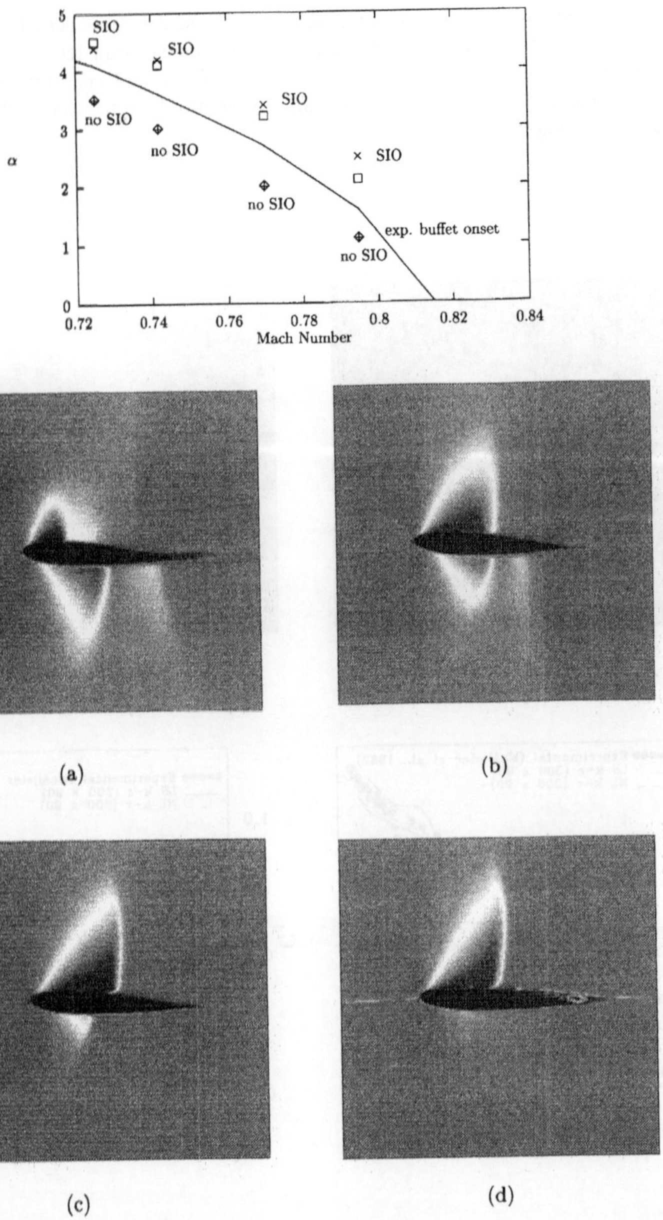

Fig. 1.10 Results (53) for the buffet onset (angle of incidence versus Mach number) for a transonic turbulent flow around a NACA-0012 aerofoil using a Godunov-type method (23) in conjunction with an implicit-unfactored solver (29), (46). The solutions obtained by the one-equation SA model (52) (crosses) and a non-linear eddy-viscosity model (47) (squares), are compared with the experimental results from (54); SIO stands for 'shock-induced oscillations'. The flow field at different time instants is also shown.

The above advocate that accurate simulation of turbulent flows can possibly be performed at scales much larger than the Kolmogorov scale. Independent research studies (1)–(3), (21), have shown that LES of turbulent flows can also be performed on coarse grids without using a SGS model, if high-resolution monotone methods are employed for solving the flow equations. In this case the numerical solution of the Navier–Stokes equations is filtered through the numerical scheme and the accuracy of the simulation relies entirely on the dissipation and dispersion properties of the non-linear monotone advection method.

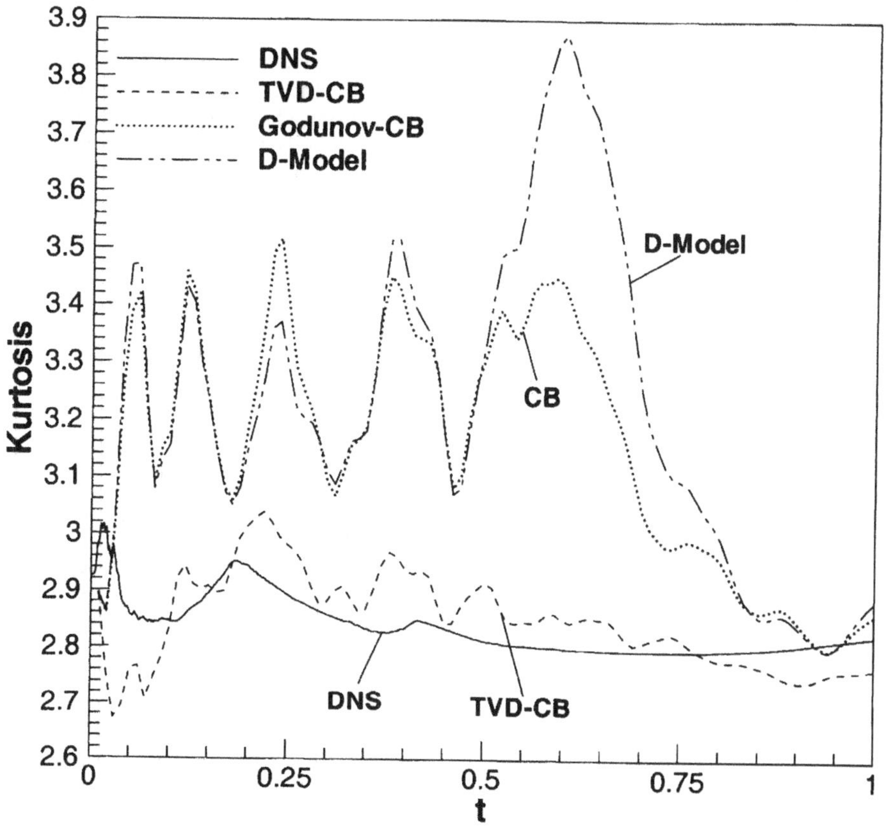

Fig. 1.11 Kurtosis distributions for the Burgers' turbulence using different computational approaches (3) (see the text for more details). The best results are obtained by the TVD-CB Godunov-type scheme without using a SGS model.

In (3), this approach is referred to as 'VLES' or 'Very Large Eddy Simulation'. In this direction, particular efforts should be spent in developing and investigating non-linear monotone schemes which would lead to accurate representation of the large energetic scales. An example from our recent work on VLES is shown in Fig. 1.11. We have performed simulations in the context of Burgers' turbulence[4] (**57**), (**58**) by employing different Godunov-type methods with and without utilizing a SGS model. In Fig. 1.11, we show results from high-order statistics as obtained by the VLES computations using the following three approaches:

(i) The Godunov-type CB scheme without a SGS model (this solution is labelled 'CB');
(ii) A recently developed TVD version of the CB scheme (**3**) without a SGS model (this solution is labelled 'TVD-CB');
(iii) The CB scheme in conjunction with the modified version (**59**) of the dynamic SGS model (**60**) (this solution is labelled 'D-Model').

The above solutions are compared with the results obtained by DNS of the Burgers' turbulence, using a very fine grid and a small time step (**3**). The computations reveal that the best results are obtained by the TVD-CB scheme without using a SGS. Moreover, the addition of the SGS model on the CB scheme does not lead, overall, to better results. In (**3**), we have also performed similar investigations for mixing layer flows. Currently, we are also conducting studies regarding the use of the TVD-CB scheme and other Godunov-type methods as an 'implicit modelling' approach in complex transitional and turbulent flows of engineering interest.

1.5. Concluding remarks

We discussed a number of computational approaches for simulating steady and time-dependent, laminar and turbulent, as well as incompressible and compressible flows. We highlighted, in particular, the benefits with respect to numerical accuracy that one can derive by using non-linear monotone methods such as Godunov-type methods. Intensive research using these methods for accurately predicting shock waves and other gasdynamic phenomena, has been conducted for over three decades. Research to fully exploit the properties of these methods in the simulation of transitional and turbulent flows, is still in its infancy. The preliminary indications are very encouraging, but significant efforts still need to be spent in order to understand the dissipation and dispersion behaviour of these methods in the above flows.

Relevant to the above understanding is also the issue of numerical artifacts produced by computational methods (**61**), (**62**). The differential equations are represented by difference equations and thus spurious solutions due to the numerical scheme, time step, initial and boundary conditions, and time interval for which the calculation proceeds, may be introduced. From the perspective of numerical analysis, an understanding of the occurrence of spurious solutions is buried in the details of the truncation error. In the past, phenomena of stable and unstable multiple solutions and spurious steady state numerical solutions occurring below and above the linearized stability limit of a numerical scheme were observed (**63**). Further research using the Navier–Stokes equations (**35**) has also shown that bifurcations to and from spurious asymptotic solutions are not only highly scheme and problem dependent, but also initial data and boundary condition dependent. Therefore, simulations of complex flow phenomena such as turbulence should always be considered bearing in mind the aforementioned uncertainties.

To achieve high-accuracy in under-resolved simulations of flows of engineering interest is a major challenge. This can only be done by developing high-order methods which satisfy the

4 The Burgers' equation can be considered as an one-dimensional analog to the Navier–Stokes equations, though they lead to different energy spectra.

properties mentioned in Section 1.2. The combination of these methods with advanced acceleration algorithms such as the dynamically-adaptive multigrid can possibly provide the desired accuracy in short computing times thus making complex turbulent flow computations affordable in an industrial design environment.

1.6 References

(1) **Oran, E. S.** and **Boris, J. P.** (1993) *J. Comp. Phys.*, **7**, 523–533.

(2) **Margolin, L. G., Smolarkiewicz, P. K.**, and **Sorbjan, Z.** (1999) *Physica D*, **133**, 171–178.

(3) **Drikakis, D.** (2001) (to appear) Proceedings of *ECCOMAS CFD Conference*.

(4) **Godunov, S. K.** (1959) *Mat. Sb.*, **47**, 357–393.

(5) **Boris, J. P.** and **Book, D. L.** (1976) *J. Comp. Phys.*, **20**, 397–431.

(6) **Van Leer, B.** (1979) *J. Comp. Phys.*, **32**, 101–136.

(7) **Roe, P. L.** (1981) *J. Comp. Phys.*, **43**, 357–372.

(8) **Osher, S.** and **Solomon, F.** (1982) *Math. Comp.*, **38**, 339–374.

(9) **Harten, A.** (1983) *J. Comp. Phys.*, **49**, 357–393.

(10) **Colella, P.** and **Woodward, P. R.** (1984) *J. Comp. Phys.*, **54**, 174–201.

(11) **Ben-Artzi, M.** and **Falcovitz, J.** (1984) *J. Comp. Phys.*, **55**, 1–32.

(12) **Sweby, P. K.** (1984) *Siam J. Numer. Analysis*, **21**, 995–1011.

(13) **Yee, H. C.** (1987) *J. Comp. Phys.*, **68**, 151–179.

(14) **Colella, P.** (1990) *J. Comp. Phys.*, **87**, 171–200.

(15) **Billet, S. J.** and **Toro, E. F.** (1997) *J. Comp. Phys.*, **130**, 1–24.

(16) **Zoltak, J.** and **Drikakis, D.** (1998) *Comp. Methods Appl. Mech. Engng*, **162**, Nos. 1–4, 165–185.

(17) **Toro, E. F.**, (1999) *Riemann Solvers and Numerical Methods for Fluid Dynamics*, Second edition, Springer.

(18) **Bell, J. B., Colella, P.**, and **Glaz, H. M.** (1989) *J. Comp. Phys.*, **85**, 257–283.

(19) **Drikakis, D., Govatsos, P. A.**, and **Papantonis, D. E.** (1994) *Int. J. Numer. Methods Fluids*, **19**, 667–685.

(20) **Drikakis, D.** (1996) *Advances Engng Software*, **26**, 111–119.

(21) **Drikakis, D.** (2001) *Godunov Methods: Theory and Applications*, (Edited by E. F. Toro), Kluwer Academic Publishers, pp. 263–283.

(22) **Drikakis, D., Iliev, O.**, and **Vassileva, D. P.** (1998) *J. Comp. Phys.*, **146**, 301–321.

(23) **Barakos, G.** and **Drikakis, D.** (1998) *Int. J. Numer. Methods Fluids*, **28**, 73–94.

(24) **Shu, C-W** and **Osher, S.** (1988) *J. Comp. Phys.*, **77**, 439–471.

(25) **Canuto, C., Hussaini, M. Y., Quarteroni, A.**, and **Zang, T. A.** (1988) *Spectral Methods in Fluid Dynamics*, Springer-Verlag.

(26) **Zang, T. A.** (1991) *Appl. Numer. Math.*, **7**, 27.

(27) **Drikakis, D., Ofengeim, D., Timofeev, E.**, and **Voinovich, P.** (1997) *J. Fluids Structures*, **11**, 665–691.

(28) **Ofengeim, D.** and **Drikakis, D.** (1997) *Shock Waves*, **7**, 305–317.

(29) **Barakos, G.** and **Drikakis, D.** (1999) *Comps Fluids*, **28**, 899–921.

(30) **Barakos, G.** and **Drikakis, D.** (2000) to appear in *AIAA J.*, **38**, January.

(31) **Chorin, A. J.** (1967) *J. Comp. Phys.*, **2**, 12–26.

(32) **Merkle, C. L.** and **Athavale, M.** (1987) *AIAA* Paper, 87–1137.

(33) **Soh, W. Y.** and **Goodrich, J. W.** (1988) *J. Comp. Phys.*, **79**, 113–134.

(34) **Rogers, S. E., Kwak, D.**, and **Kiris, C.** (1991) *AIAA J.* **29**, 603–610.

(35) **Drikakis, D.** (1997) *Phys. Fluids*, **9**, 76–87.

(36) **Mallinger, F.** and **Drikakis, D.** (2001) Proceedings of the *Turbulence Shear Flow and Phenomena – TSFP2 Conference*, Stockholm, Sweden.
(37) **Shapira, M., Degani, D.**, and **Weihs, D.** (1990) *Comps Fluids*, **18**, 239–258.
(38) **Drikakis, D.** (1995) *Parallel Computational Fluid Dynamics: Implementation and Results using Parallel Computers*, (Edited by A. Ecer, J. Periaux, N. Satofuka, and S. Taylor), Elsevier Science B.V., 191–198.
(39) **Harten, A., Engquist, B., Osher, S.**, and **Chakravarthy, S. R.** (1987) *J. Comp. Phys.*, **71**, 231–303.
(40) **Jiang, G-S.** and **Shu, C-W.** 1996) *J. Comp. Phys.*, **126**, 202–228.
(41) **Lax, P. D.** (1994) *Comm. Pure Appl. Math.*, **VII**, 159–193.
(42) **Rusanov, V. V.** (1961) *J. Comp. Math. Phys.* USSR, **1**, 267–279.
(43) **Jameson, A.** (1983) *Appl. Math. Comp.*, **13**, 327–356.
(44) **Drikakis, D., Iliev**, and **Vassileva, D. P.** (2000) *J. Comp. Phys.*, **165**, 566–591.
(45) **Chakravarthy, S. R.** (1988) VKI Lecture Series, *Computational Fluid Dynamics*, 1988–05.
(46) **Drikakis, D.** and **Durst, F.** (1994) *Int. J. Numer. Methods Fluids*, **18**, 385–413.
(47) **Craft, T. J., Launder, B. E.**, and **Suga, K.** (1996) *Int. J. Heat Fluid Flow*, **17**, 108–115.
(48) **Barakos, G.** and **Drikakis, D.** (2000) *Phil. Trans. R. Soc. Lond. Ser A*, **358**, 3279–3291.
(49) **Drikakis, D.** and **Barakos, G.** (2000) CD-Rom Proceedings of the *ECCOMAS Conference*.
(50) **McAlister, K. W., Pucci, S. L., McCroskey, W. J.**, and **Carr, L. W.** (1982) NASA TM-84245, September.
(51) **Launder, B. E.** and **Sharma, B. I.** (1974) Letters in *Heat Mass Transfer*, **1**, 131–138.
(52) **Spalart, P. R.** and **Allmaras, S. R.** (1992) *AIAA* Paper 92-0439.
(53) **Barakos, G.** and **Drikakis, D.** (2000) *Int. J. Heat Fluid Flow*, **21**, 620–626.
(54) **McDevitt, J. B.** and **Okuno, A. F.** (1985) NASA-TP-2485, NASA Ames.
(55) **Muti Lin, J. C.** and **Pauley, L. L.** (1996) *AIAA J.*, **34**, 1570–1576.
(56) **Kolmogorov, A. N.** (1962) *J. Fluid Mech.*, **13**, 82.
(57) **Kraichnan, R.H.** (1968) *Phys. Fluids*, **11**, 265–277.
(58) **Hosokawa, H.** and **Yamamoto, K.** (1975) *J. Stat. Phys.*, **13**, 245–272.
(59) **Lilly, D.K.** (1992) *Phys. Fluids*, **4**, 633–635.
(60) **Germano, M., Piomelli, U., Moin, P.**, and **Cabot, W. H.** (1991) *Phys. Fluids*, **3**.
(61) **Drikakis, D.** and **Smolarkiewicz, P. K.** (2000) *J. Comp. Phys.*, submitted.
(62) **Drikakis, D., Margolin, L.**, and **Smolarkiewicz, P. K.** (2001) Proceedings of *ICFD Conference*, (to appear).
(63) **Yee, H. C., Sweby, P. K.**, and **Griffiths, D. E.** (1990) NASA TM 102820, April.

2

Detection of Multiple Solutions using a Mid-cell Back Substitution Technique Applied to Computational Fluid Dynamics

S R Kendall and **H V Rao**

Synopsis

Computational models for fluid flow based on the Navier–Stokes equations for compressible fluids lead to numerical procedures requiring the solution of simultaneous non-linear algebraic equations. These give rise to the possibility of multiple solutions, and hence there is a need to monitor convergence towards a physically meaningful flow field.

The number of possible solutions, which may arise, is examined and a mid-cell back substitution technique (MCBST) is developed to detect and avoid convergence towards apparently spurious solutions.

The MCBST was used successfully for flow modelling in micron-sized flow passages, and was found to be particularly useful in the early stages of computation, optimizing the speed of convergence.

2.1 Notation

D_{eq}	Equivalent diameter	m
E_j	Represents the error resulting from the back substitution of the mid-cell (y_{im}) values into the original finite difference flow equations	
H	Passage depth	m
H_2O	Abbreviation used for de-ionized water	
L_{dev}	Flow developing length	m
L_{th}	Passage length	m
m	Maximum grid mesh number in the y–direction	
n	Maximum grid mesh number in the x–direction	
N_2	Abbreviation for nitrogen gas	

p	Pressure	N/m^2
R_e	Reynolds number	
u	Velocity component in the x–direction	m
v	Velocity component in the y–direction	m
w	Velocity component in the z–direction	m
x	x co-ordinate	m
y	y co-ordinate	m
y_{im}	The extrapolated values for u, v, and p, at the mid-cell positions of the flow mesh	
y_{io}	The solution of the unknown values for u, v, and p, at the nodal points of the flow mesh	
z	z co-ordinate	m
α_D	Arbitrary coefficient used in determination of the developing length	
Δx	Grid size in the x–direction	m
Δy	Grid size in the y–direction	m
Δz	Grid size in the z–direction	m
ρ	Fluid density	kg/m^3
μ	Fluid viscosity	kg/ms

2.2 Introduction

The non-linear nature of the simultaneous algebraic equations encountered in the numerical solution of the Navier–Stokes equations governing the flow of fluid, gives rise to the possibility of multiple solutions (1). However, only one of these numerical solutions will be a true representation of the actual physical flow field. It is, hence, necessary to monitor the convergence of the solution in order to detect and avoid physically meaningless solutions.

In this chapter, the number of possible numerical solutions which may arise is examined using an alternative approach, which may also be confirmed with the more general technique known as Bezout's Theorem (2). A mid-cell back substitution technique is developed for monitoring solution convergence.

The convergence technique developed here was originally intended for the computation of the flow fields in micron-sized passages. However, the technique may also be applied to general flow problems.

The computational results obtained for micron-sized passageways have been compared with corresponding experimental flow data. A brief description of the experimental equipment is also presented.

2.3 Theory

2.3.1 Occurrence of multiple solutions in numerical procedures employing the Navier–Stokes equations

The numerical solution of the Navier–Stokes equations based upon the finite difference equations, along with appropriate boundary conditions, may be assumed to result in algebraic equations up to a third degree, as follows:

$N_1 =$ number of linear algebraic equations,
$N_2 =$ number of quadratic algebraic equations,
$N_3 =$ number of cubic algebraic equations.

The procedure may readily be extended to accommodate higher than third order algebraic equations, when necessary. The total number of algebraic equations, N, is given by:

$$N = (N_1 + N_2 + N_3)$$

The above equations are written in terms of the unknown values of u, v, and p at various nodal points for a steady state problem. It is shown in Appendix A, that the total possible sets of solutions is given by:

$$N_{set} = 2^{N_2} \times 3^{N_3} \tag{2.1}$$

It is possible that some of these roots may be imaginary.

2.3.2 Mid-cell back substitution technique applied to two-dimensional flow

Let the finite difference equations for the unknown u, v, and p values at the nodal points within a two-dimensional flow field be represented by the following 'N' equations:

$$f_j(y_1, y_2, \ldots\ldots\ldots\ldots, y_N) = 0, \qquad \text{for } j = 1, 2, \ldots\ldots\ldots\ldots, N \tag{2.2}$$

where y_i, $i = 1, 2, \ldots\ldots\ldots\ldots, N$ are the unknown values of u, v, and p at the nodal points.

If y_{io}, $i = 1, 2, \ldots\ldots\ldots\ldots, N$ is one of the solutions to the 'N' algebraic equations, then

$$f_j(y_{1o}, y_{2o}, \ldots\ldots\ldots\ldots, y_{No}) = 0 \qquad \text{for } j = 1, 2, \ldots\ldots\ldots\ldots, N \tag{2.3}$$

The solution of the unknown values for u, v, and p are represented by y_{io} at various grid nodal points shown in Fig. 2.1, and these y_{io} values may be used to extrapolate mid-cell values denoted by the symbol y_{im}, for $i = 1, 2, \ldots, N$.

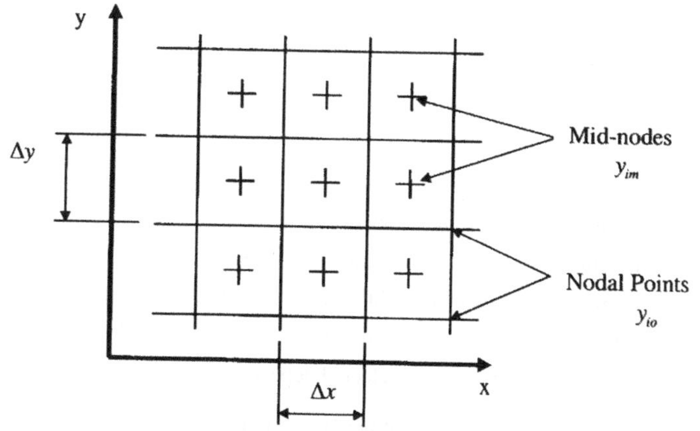

Fig. 2.1 Two-dimensional flow field depicting grid system and mid-nodes

Using Taylor's expansion (3):

$$y_{im} \approx y_{io} + \frac{\partial y_i}{\partial x}\left(\frac{\Delta x}{2}\right) + \frac{\partial y_i}{\partial y}\left(\frac{\Delta y}{2}\right)$$

$$\therefore \; y_{im} = y_{io} + \delta_i$$

where,

$$\delta_i \equiv \left(y_{im} - y_{io}\right) \tag{2.4}$$

Substitution of y_{im}, $i = 1, 2, \ldots, N$, into the functions, f_j, $j = 1, 2, \ldots, N$ defined in equation (2.2), provides magnitudes of f_j, which are equated to E_j, as defined below:

$$E_j = f_j\left(y_{1m}, y_{2m}, \ldots, y_{Nm}\right), \qquad j = 1, 2, \ldots, N \tag{2.5}$$

E_j, $j = 1, 2, \ldots, N$, represent the 'error' resulting from the back substitution of the mid-node (y_{im}) values into equation (2.2).

E_j may alternatively be expressed, thus:

$$E_j = f_j\left(y_{1m}, y_{2m}, \ldots, y_{Nm}\right) - f_j\left(y_{1o}, y_{2o}, \ldots, y_{No}\right)$$

$$= \sum_{i=1}^{i=N}\left[\left(\frac{\partial f_j}{\partial y_i}\right)_{y_i = y_{io}} \delta_i\right] + \left(\text{Terms of } \delta_i \text{ raised to the power of 2 and above}\right) \tag{2.6}$$

where δ_i are defined in equation (2.4).

The partial derivatives $\left(\frac{\partial f_j}{\partial y_i}\right)_{y_i = y_{io}}$, may have magnitudes that result in the following possible inequalities:

$$E_{max} / \delta_{min} < E_o \tag{2.7}$$

$$E_{min} / \delta_{max} > E_o \tag{2.8}$$

where

δ_{min} ≡ Minimum of the absolute values of δ_i, $i = 1, 2, \ldots, N$

δ_{max} ≡ Maximum of the absolute values of δ_i, $i = 1, 2, \ldots, N$

E_{min} ≡ Minimum of the absolute values of E_j, $j = 1, 2, \ldots, N$

E_{max} ≡ Maximum of the absolute values of E_j, $j = 1, 2, \ldots, N$

E_o is a positive number suitably selected dependent upon the grid aspect ratio $\left(\Delta y / \Delta x\right)$, usually having a value between 1 and 10.

If the solutions of the y_{io} values are such that the inequality in (2.7) is valid, then the solution is considered acceptable. If inequality (2.8) is true, then the corresponding solution is rejected.

2.3.3 The principle of the mid-cell back substitution technique

If the y_{im} values were inserted into the equations represented by (2.2), they would result in an error E_j, as given in equation (2.5). Where the condition given by inequality (2.7) is satisfied, the maximum error E_{max}, will be small.

On the other hand, if the solutions for y_{io} result in E_j values, which satisfy the condition given in inequality (2.8), then this will lead to a large value of E_{max}.

In the mid-cell back substitution procedure described in Section 2.3.2, a sequence of iterates is generated that converges to the solution of the system, provided that the initial approximations 'y_{io}', derived from the inlet boundary conditions, are sufficiently close to the true solution. By monitoring the convergence of the solution using the mid-cell substitution extrapolations, any divergence of the generated errors can be detected at an early stage in the solution procedure. The initial approximations 'y_{io}', may then be adjusted within pre-determined increments and limits, and the whole solution procedure re-initiated.

Hence, the mid-cell back substitution procedure is a means of avoiding solutions y_{io}, which generate large values of $\left(\partial f_j / \partial y_i\right)$ which lead to large 'E_j' values.

2.3.4 The two-dimensional flow model

The two-dimensional flow model for micron-sized passageways was developed based upon the finite difference form of the Navier–Stokes equations. In order to accommodate the large aspect ratios encountered with the flow field grids for micro-passage flows, a 'block marching procedure' was developed to minimize the number of simultaneous equations to be solved (**4**).

2.3.5 Computer resources

The computer used extensively in the solution of the numerical flow model was a Fujitsu MCCVPX240/10 supercomputer located at the University of Manchester, Oxford Road, Manchester, UK. Access to this computing system was confined to 12-hour slots, once or twice a week, depending on overall demand.

2.4 Experimental study of flow through micron-sized passages

In order to establish the validity of the theoretical models, an experimental investigation of the flow characteristics in micron-sized passageways was undertaken. The design and construction of micro-passages, pressure transducers, and a positive displacement flow metering test-rig was necessary.

2.4.1 Construction of the micron-sized test passages

The micron-sized test passages were designed to emulate two-dimensional flow to achieve the best comparison possible between experimental results and those obtained from a two-dimensional computational model. The passage widths were made at least 100 times greater than their respective depths in order to accomplish this.

The passages were designed for a simple method of manufacture based on the use of polyimide thin film spacers, (see Fig. 2.2).

The surface roughness of the Perspex used in the fabrication of the micro-passages was mapped using a 'DEKTAK' surface tester, which indicated a surface roughness averaging out to no more than 0.1 μm.

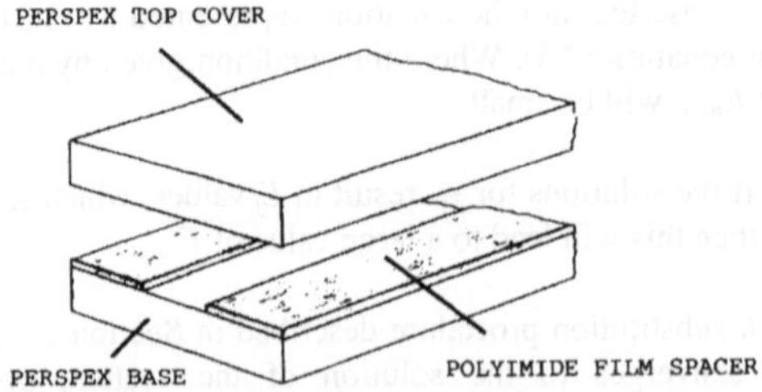

Fig. 2.2 Micro-passage construction

2.4.2 Flow parameters measured

The two most useful flow parameters which can be measured directly and best describe the nature of fluid flow, are (a) pressure, and (b) flow rate.

Due to the very small passage dimensions involved, specially designed pressure transducers were required to measure the pressure differentials. The mass flow rate can be precisely metered and controlled using a positive displacement piston.

2.4.3 Pressure transducers

In the absence of commercially available products for pressure measurement in micron-sized flow systems, investigations indicated that a diaphragm/strain gauge design provided the best means of measuring pressure differentials along a micron-sized passage. The main influencing factors in this decision were, the availability of diaphragm strain gauges making it possible to build pressure transducers that complement the size of the micro-passages.

Figure 2.3 shows a special purpose diaphragm strain gauge designed specifically for use in pressure transducers (5). The full-bridge patterns shown are designed to take maximum advantage of the strain distribution on a rigidly clamped diaphragm of uniform thickness.

The pressure transducer design, incorporating diaphragm strain gauges, can be seen in Fig. 2.4, and is specifically intended for the measurement of pressure differentials.

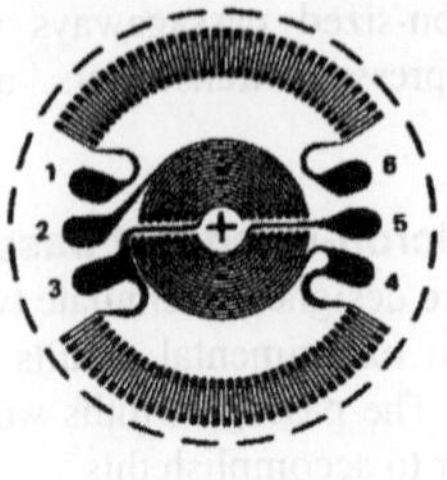

Fig. 2.3 Diaphragm strain gauge

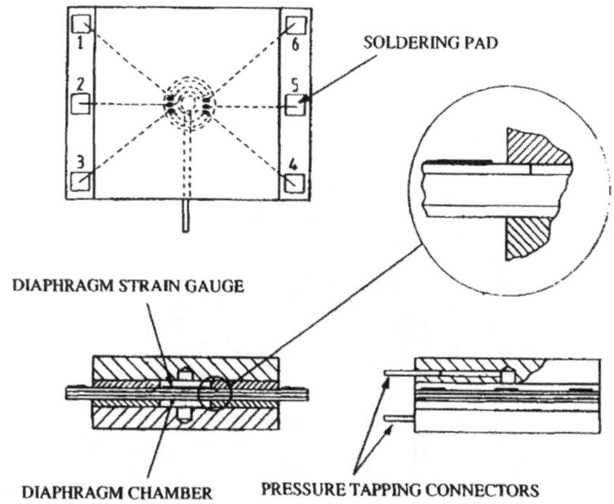

Fig. 2.4 Pressure transducer assembly

2.4.4 Delivering precise steady fluid flows

The most important feature of the main test-rig is that it supplies a very precise steady flow of fluid to the micro-passages. The required precision and consistency for the flow rates was accomplished using positive displacement piston/cylinder arrangements (refer to Fig. 2.5).

The substantial difference in viscosity and density between nitrogen and de-ionized water precluded the use of a single flow assembly for supplying both these fluids. Hence, two separate assemblies were designed. The larger assembly shown on the left-hand side of Fig. 2.5 is for delivering nitrogen gas, while the smaller arrangement on the right supplies de-ionized water.

2.4.5 Performance of the experimental test-rig

The test-rig worked well as a whole, proving very reliable in delivering a precise and constant flow rate of both nitrogen gas and de-ionized water to an accuracy of 0.00116 per cent.

The pressure transducers worked particularly well, demonstrating good linearity in their outputs and responding to pressure changes quickly in accordance with their designed stabilizing times.

2.5 Results

2.5.1 Results obtained using an experimental test-rig

Experiments were conducted with various passage depths measured in micro-metres, two fluids, and a variety of inlet flow rates as determined by the positive displacement of a piston.

Fig. 2.6 shows a schematic layout of the unique experimental test-rig employed in the validation of the computational results.

Fig. 2.5 Piston/cylinder assemblies

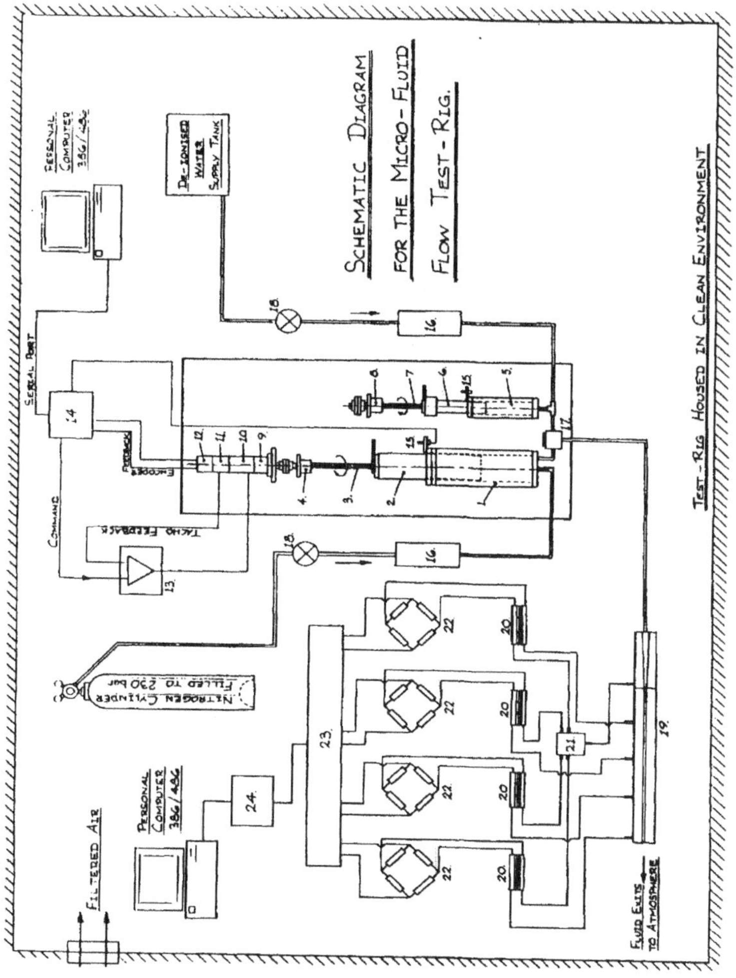

Fig. 2.6 Schematic drawing of test-rig and ancillary equipment

Key to accompany Fig. 2.6

1. Phosphor Bronze Cylinder (Nitrogen), I.D. = 80 mm
2. Stainless Steel Piston (Nitrogen), O.D. = 80 mm
3. 16 mm Diameter × 2 mm Lead Precision Ball Screw
4. Bearing Support Unit for 16mm Diameter Ball Screw
5. Phosphor Bronze Cylinder (Water), I.D. = 20 mm
6. Stainless Steel Piston (Water), O.D. = 20 mm
7. 10 mm Diameter × 1 mm Lead Precision Ball Screw
8. Bearing Support Unit for 10 mm Diameter Ball Screw
9. 43:1 ratio Gearhead
10. D.C. Servo Motor
11. Tachogenerator
12. Shaft Encoder
13. Servo Amplifier
14. Motor Controller
15. Limit Switch
16. 0.5 μm In-line Filter (stainless steel)
17. 3-way Ball Valve (stainless steel)
18. In-line Ball Valve (stainless steel)
19. Micron-sized Passage
20. Pressure Transducer (diaphragm strain gauge design)
21. Micro-Manifold
22. Strain Indicator, configured as a full bridge
23. Simultaneous Sample and Hold Board
24. Analogue to Digital Converter

2.5.2 Results from the computational model

The flow fields, utilizing the MCBST, were computed for the micro-passages corresponding with the experimental studies undertaken.

From experience, the permissible error in the finite difference equations arising from back substitution can be related to the size of the flow field mesh.

A satisfactory back substitution error value is governed by the size of the flow field mesh employed.

A weighted correction technique was employed within the iterative solution procedure. Using the dependent variable u, as an example of the dependent variables u, v, and p, the weighted correction in u is defined as:

$$\text{Weighted correction } (u) = \frac{\text{accumulative change in } u \text{ due to solution procedure}}{\text{accumulative } u \text{ terms}}$$

or

$$\text{Weighted correction } (u) = \frac{\sum\left\{ (u_o \times u') \Big/ (u_o + u') \right\}}{\sum (u_o)} \tag{2.9}$$

where, u_o is the initial estimate for the u-velocity, and u' is the change in u_o as produced by the iterative solution procedure. If the weighted correction in either u, v, and p, is greater than desired permissible levels, then new initial estimates are obtained using the following equation, and the whole iterative solution procedure resumed:

$$u_o = u_o + u' \tag{2.10}$$

To promote stability of a solution, especially in the early stages of convergence, a relaxation factor can be applied to the dependent variables u, v, and p, by introducing a relaxation coefficient into equation (2.10) in the following manner:

$$u_o = u_o + \alpha_R \times u' \tag{2.11}$$

The corrections in the dependent variables were determined using the LU decomposition procedure (**6**), and the relaxation coefficient α_R was usually set between 0 and 1 (i.e. underrelaxation). The optimum relaxation coefficient can vary depending upon the type of fluid, the number of nodal points on the finite difference mesh, and the mesh spacing itself. Keeping the relaxation coefficient constant throughout the entire convergence procedure was found to be inefficient. Therefore, a system which monitors convergence was incorporated which gradually adjusts the relaxation coefficient in accordance with the convergence trends exhibited by the dependent variables.

Typical results for fluid flow in micron-sized passages can be exemplified by the relationship of the fluid's centre-line u-velocity compared against the length of passage through which it flows. Figure 2.7 shows how this value for the centre-line u-velocity initially peaks following inlet to the passageway and then steady's out.

It is this type of flow behaviour which benefited from the inclusion of the MCBST because it enhanced the likelihood of the numerical solution converging.

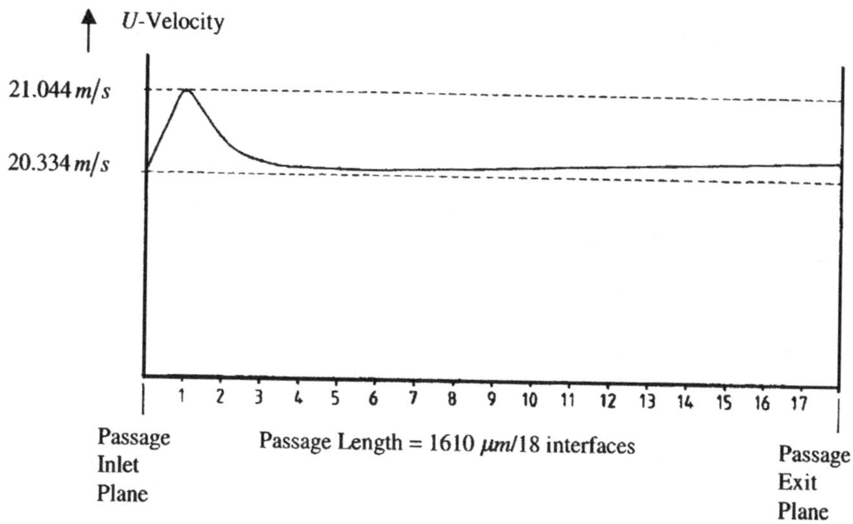

Fig. 2.7 Variation in the centre-line *u*-velocity along the length of the passage

2.5.3 Results obtained using a commercial CFD package

As part of the analysis program, a commercially available package was also utilized in analysing the flow of nitrogen gas through a 55 μm deep rectangular passage. The commercial package chosen adopts the pressure correction equation (7) in its numerical procedure, and sets up a system of equations for the global solution of the flow field.

In making a comparison with the 'block marching system' that was devised originally for the numerical analysis of micro-passage flows (see Section 2.3.4), the centre-line *u*-velocity was the prominent feature chosen for comparison purposes.

The results from the commercial program compared very favourably with the centre-line *u*-velocity from the 'block marching system' shown in Fig. 2.7, in fact within 3 per cent overall. This shows that the 'block marching procedure' incorporating the MCBST is a viable CFD technique.

2.5.4 Comparison of results using the friction factor variable

Since small values of Reynolds Number may be affiliated with micro-passage flows, the relationship between friction factor *f*, and the Reynolds Number in laminar steady flows through micron-sized passages was examined.

Friction factors were obtained for a range of micro-passage flow cases, using both computational and experimentally determined pressure gradients (*dp/dx*). The Fanning friction factor (8) is defined by the following equation:

$$f = \frac{(dp/dx)\, D_{eq}}{2\, \rho_m\, u_m^2} \tag{2.12}$$

The friction factors determined from the computational and experimental results are compared with theoretically predicted values for one-dimensional incompressible laminar steady flow (9), given by the following equation:

$$f = \frac{24}{\mathrm{Re}} \tag{2.13}$$

The results for friction factors from the experimental work with micro-passages, the corresponding computational results based on two-dimensional laminar steady flow, and traditional correlations for one-dimensional fully developed laminar flow between parallel plates, are presented in Table 2.1.

Table 2.1 Comparison between theoretical, computational and experimentally found friction factors in micron-sized passages

Passage depth and type of Fluid, respectively	Reynolds Number Re	Friction factor f calculated using the equation $f = \frac{24}{\mathrm{Re}}$	Friction factor f, calculated from Computational pressure variations along micro-passages	Friction factor f, calculated from Experimental pressure variations along micro-passages
30 μm De-ionized water	13.2	1.818	2.349	2.463
30 μm Nitrogen gas	49.3	0.487	0.643	0.706
55 μm De-ionized water	82.4	0.291	0.379	0.405
55 μm Nitrogen gas	109.6	0.219	0.239	0.239 *
1 mm De-ionized water	800	0.03	0.046	-

The comparison of friction factors determined from computational and experimental results exhibit a good correlation. The theoretically predicted traditional values are, however, consistently lower than the other two.

An error analysis for the experimentally determined friction factors has been carried out. In the specific case corresponding to the value marked * in Table 2.1, the probable error in the friction factor was found to be ± 0.0207.

2.5.5 Developing length and the variation of $\partial u/\partial x$

The variation of centre-line $\partial u/\partial x$ along the passage length is a useful means of determining the developing length of the flow.

The relationship for developing length in laminar flow can be written:

$$\frac{L_{dev}}{D_{eq}} = \alpha_D \, \text{Re} \tag{2.14}$$

where α_D is an arbitrary coefficient, $D_{eq} = 2H$, and $\text{Re} = \dfrac{\rho u_{max}(2H)}{\mu}$.

Table 2.2 shows the range of magnitudes for the coefficient α_D when considering macro-flows in circular pipes and flow between parallel plates/rectangular passages.

Table 2.2 Values for α_D in Macro–flow systems

Macro-flow system	α_D	Reference
Circular pipes	0.058 to 0.06	**(10), (11)**
Parallel plates/rectangular passages	0.0065	**(12)**

The results from the computational analysis were used to establish developing lengths using the following exponential projection method, to ensure consistency.

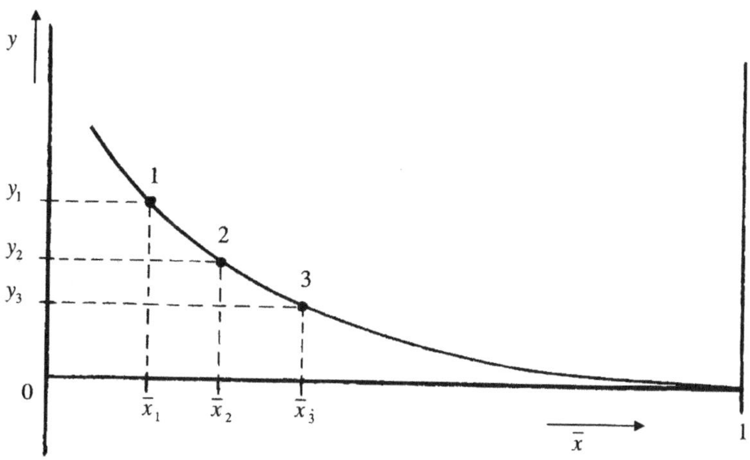

Fig. 2.8 Determining developing lengths using a exponential projection method

Using the exponential expression:

$$y = a\, e^{-b/(1-\bar{x})} \tag{2.15}$$

where a and b are arbitrary constants.

Applying equation (2.15) between points 1 and 2 in Fig. 2.8:

$$\left(\frac{y_2}{y_1}\right) = e^{-b\left(1/(1-\bar{x}_2) - 1/(1-\bar{x}_1)\right)}$$

$$\therefore\ b' = \frac{\ln(y_2/y_1)}{\left(\frac{1}{(1-\bar{x}_1)} - \frac{1}{(1-\bar{x}_2)}\right)} \qquad \text{and} \qquad a' = \left(y_1 \Big/ e^{-b'/(1-\bar{x}_1)} \right)$$

Similarly, applying equation (2.15) between the points 1 and 3, then 2 and 3 in Fig. 2.8, enables the determination of the constants a'', b'', and a''', b'''. For each pair of points a value of $\bar{x}^*$, corresponding to a y value of 0.01, is obtained as follows:

$$\bar{x}^* = 1 + \frac{b}{\ln\left(\frac{0.01}{a}\right)} \tag{2.16}$$

The results are tabulated as follows:

Table 2.3 Arbitrary constants combinations for the exponential projection procedure

Points (Refer to Fig. 2.6)	Constants		$\bar{x}^*$
1 and 2	a'	b'	$\bar{x}_1^*$
1 and 3	a''	b''	$\bar{x}_2^*$
2 and 3	a'''	b'''	$\bar{x}_{3^*}$
Mean			$\frac{\bar{x}_1^* + \bar{x}_2^* + \bar{x}_3^*}{3}$

The mean value of $\bar{x}^*$ in Table 2.3 is then used to represent the position along the $\bar{x}$-axis in Fig. 2.8 which corresponds to a y value that has reached 99 per cent of it's final value. Hence,

$$L_{dev} = \bar{x}^* \times L_{th} \tag{2.17}$$

With the computational data obtained for various micro-passage cases, and using equation (2.14) the following table of α_D values has been prepared.

Table 2.4 The relationship between Reynolds number and developing length in micron-sized passages

Passage depth	Fluid	Re	α_D
30 μm	De-ionized water	13.2	0.057
30 μm	Nitrogen gas	49.3	0.049
55 μm	De-ionized water	82.4	0.029
55 μm	Nitrogen gas	109.6	0.025
1 mm	De-ionized water	800	0.018

Table 2.4 shows that for very small Reynolds numbers the α_D coefficient appears to attain higher values. It is possible that the relationship between α_D and Re is as shown in Fig. 2.9.

In Fig. 2.9, all the α_D values tabulated in Table 2.4 fall within the range denoted by micro-continuum flow.

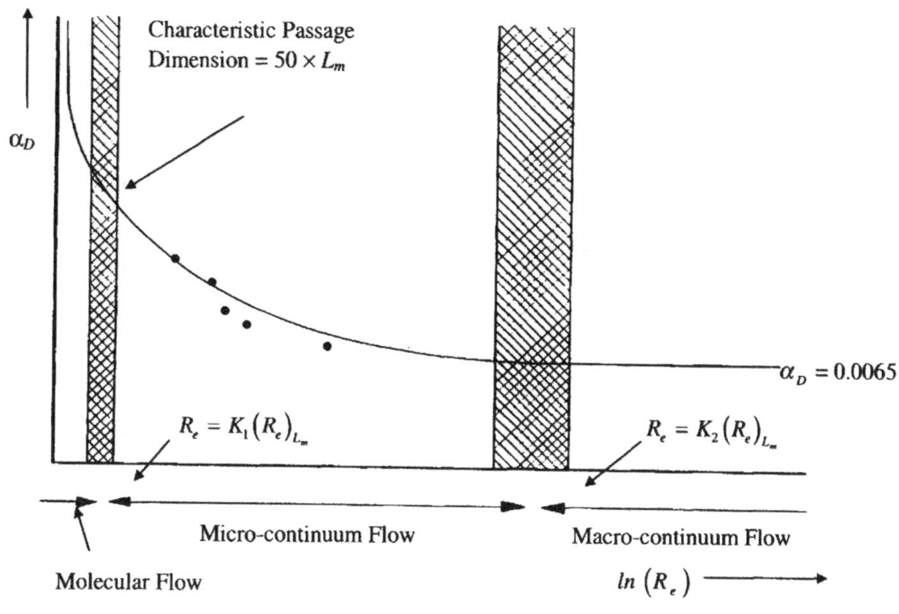

Key for Fig. 2.7

$(R_e)_{L_m}$ is the Reynolds Number for a passage having a depth equal to the molecular mean free path of the fluid.

$K_1 = 50$

$K_2 \approx 50{,}000$

Fig. 2.7 Reynolds Number and α_D relationship

2.5.6 Modelling compressibility

A practical method of establishing the validity of the theoretical model involves the comparison between the magnitudes of pressure gradient $(\partial p/\partial x)$, determined from the experimental and computational analysis of corresponding flow cases. Comparisons showed a good agreement between experimental and computational results, with the largest difference amounting to 8.9 per cent, and the smallest to 0.13 per cent, refer to Table 2.5.

Table 2.6 gives the identification codes used to denote specific passage dimensions and flow conditions, all of which have been analysed using both the computational model and the experimental test-rig.

2.5.7 Convergence of the computational procedure

The convergence of the numerical solution utilizing a 55 μm deep passageway and Nitrogen gas as the working fluid (hence the passage designation N55a), is shown in Fig. 2.10.

It can be observed from the figure that convergence was a little erratic at first, indicating the value of the mid-cell back substitution procedure in helping to avoid divergence of the solution in the early stages of computation.

Table 2.5 Comparison between computational and experimental $\partial p/\partial x$ values

Computational Simulation Identification (Refer to Table 2.6)	Computational Results for $\partial p/\partial x$ (Pa/mm)	Experimental Results for $\partial p/\partial x$ (Pa/mm)	Difference between Computational and Experimental Results (%)
N30a	- 785	- 750	4.7
N30b	- 2140	- 2350	8.9
N55a	- 988	- 1000	1.2
N55b	- 1598	- 1600	0.13
W30a	- 1428	- 1550	7.9
W30b	- 2384	- 2500	4.6
W55a	- 1192	- 1200	0.7
W55b	- 2435	- 2600	6.3

Table 2.6 Identification codes used to denote the flow parameters

Passage Identification	Fluid	Passage Depth (μm)	Passage Length (μm)	Passage Inlet Pressure (bar)	Maximum u-velocity (m/s)	Maximum v-velocity (m/s)
N30a	N_2	30	901	1.85–1.89	5.026	0.352
N30b	N_2	30	901	2.72–2.95	13.404	0.938
N55a	N_2	55	1610	2.0	20.334	1.423
N55b	N_2	55	1651	3.0	34.043	2.383
W30a	H_2O	30	901	2.57	0.157	0.011
W30b	H_2O	30	901	3.56	0.262	0.018
W55a	H_2O	55	1651	2.0	0.442	0.031
W55b	H_2O	55	1651	3.25	0.892	0.062

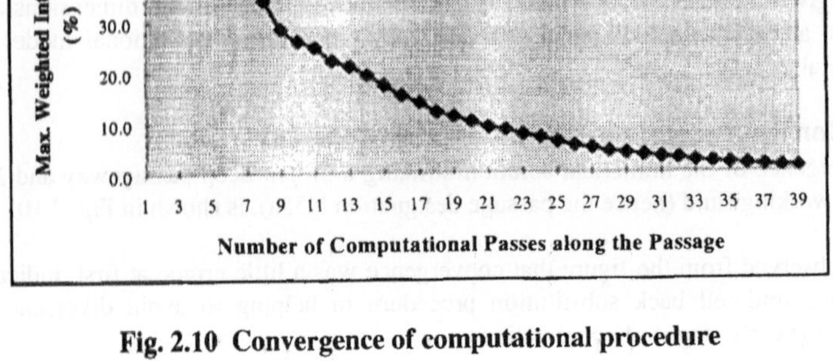

Fig. 2.10 Convergence of computational procedure

2.6 Conclusions

The mid-cell back substitution procedure was found particularly useful in the early stages of computation, where early detection of spurious solutions can avoid unnecessary computing time. The simulation of fluid flow in passages < 20 μm in depth demonstrated a good deal of sensitivity to inlet boundary conditions, in addition to aspect ratios. The mid-cell back substitution technique (MCBST) avoided unnecessary pursuance of non-physical solutions.

The CFD analysis of micro-passage flows inevitably result in large aspect ratios, giving rise to a large number of equations. The 'block marching system' alleviates this problem by dividing the passage up into more manageable block lengths involving fewer simultaneous equations in their solution.

The adoption of the mid-cell back substitution procedure can be used as an alternative to the 'staggered grid' technique (**13**) whereby the velocity and pressure components are calculated at separate grid locations. The staggered grid technique requires additional computer storage space, which is not necessary with the mid-cell back substitution procedure.

The developing lengths evaluated from the two-dimensional computational model were found to be greater than values indicated in the literature. The developing length was also found to increase in a non-linear manner with lower values of Reynolds Number.

The friction factors at low Reynolds Numbers compared with traditional correlations suggest that a higher degree of sensitivity to the compressible nature of the fluid is exhibited by flows in micro-passages.

The use of the polytropic relationship between pressure and density appears to be a tenable approximation for modelling the effects of compressibility in two-dimensional laminar steady flow.

2.7 References

(1) **Kendall, S. R.** (1996) *Analysis of fluid flow through micron sized rectangular passages*, PhD thesis, Chapter 4, Section 4.7, Solution Convergence, The University of Huddersfield.

(2) **Raghavan, M.** and **Roth, B.** (1995) Solving polynomial systems for the kinetic analysis and synthesis of mechanism and robot manipulators, *Trans ASME*, **117**, 71–79.

(3) **Stroud, K. A.** (1987) *Further Engineering Mathematics*, MacMillan Education Limited, pp. 259–261, ISBN 0-333-34875-3.

(4) **Kendall, S. R.** (1996) *Analysis of Fluid Flow Through Micron Sized Rectangular Passages*, PhD thesis, Chapter 4, Section 4.6.3, The University of Huddersfield.

(5) Diaphragm Strain Gauge N2A 06 2102H-350: Resistance of 120Ω ± 1.0% and a gauge diameter of 6.0 mm. Supplier – Measurements Group UK Limited, Basingstoke.

(6) **Press, W. H., Flannery, B. P., Teukolsky, S. A.,** and **Vetterling, W. T.** (1990) *Numerical Recipes*, Cambridge University Press, pp. 31–38, ISBN 0-521-38330-7.

(7) **Patankar, S. V.** (1980) *Numerical Heat Transfer and Fluid Flow*, Hemisphere Publishing Corporation, pp. 124–126, ISBN 0-89116-522-3.

(8) **McAdams, W. H.** (1958) *Heat Transmission*, McGraw-Hill Book Company, pp. 155–159, ISBN 0-07-Y854831-1.

(9) **Sucec, J.** (1985) *Heat Transfer*, Wm. C. Brown Publishers, pp. 434–435, ISBN 0-697-00257-8.

(10) **Fox, R. W.** and **McDonald, A. T.** (1985) *Introduction to Fluid Mechanics*, Third edition, John Wiley & Sons, pp. 331–333, ISBN 0-471-82106-3.

(11) **Langhaar, H. L.** (1942) Steady flow in the transition length of a straight tube, *J. Appl. Mech.*, **64**.

(12) **Sucec, J.** (1985) *Heat Transfer*, Wm. C. Brown Publishers, pp. 425–427, ISBN 0-697-00257-8.

(13) **Patankar, S. V.** (1980) *Numerical Heat Transfer and Fluid Flow*, Hemisphere Publishing Corporation, pp. 115–126, ISBN 0-89116-522-3.

Appendix A

If y_{io}, $i = 1, 2, \ldots, N$ are possible solutions to the N equations in (2.1), then let us denote the y_{io} solution values as follows:

$$\left.\begin{array}{rcl} y_{1,o}, y_{2,o}, \ldots, y_{N_1,o} & \equiv & r_1, r_2, \ldots, r_{N_1} \\ y_{N_1+1,o}, y_{N_1+2,o}, \ldots, y_{N_1+N_2,o} & \equiv & s_1, s_2, \ldots, s_{N_2} \\ y_{N_1+N_2+1,o}, y_{N_1+N_2+2,o}, \ldots, y_{N,o} & \equiv & t_1, t_2, \ldots, t_{N_3} \end{array}\right\} \tag{2.A1}$$

Thus, $r_1, r_2, \ldots, r_{N_1}$; $s_1, s_2, \ldots, s_{N_2}$; $t_1, t_2, \ldots, t_{N_3}$ is the alternative nomenclature for the y_{io}, $i = 1, 2, \ldots, N$ values.

Also, let the equations in (2.1) be represented as follows:

N_1 Linear equations – $l_1, l_2, \ldots\ldots\ldots\ldots, l_{N_1}$
N_2 Quadratic equations – $q_1, q_2, \ldots\ldots\ldots\ldots, q_{N_2}$
N_3 Cubic equations – $c_1, c_2, \ldots\ldots\ldots\ldots, c_{N_3}$

Step A

Assume that $r_1, r_2, \ldots, r_{N_1}$, $t_1, t_2, \ldots, t_{N_3}$ and $s_2, s_3, \ldots, s_{N_2}$ are known. Substituting these values into the quadratic equation 'q_1' gives us:

$$F(s_1) = 0 \tag{2.A2}$$

where, $F(s_1)$ will be a quadratic expression of s_1. Hence, in principle, we get two roots of s_1 from (2.3), which are represented as s_1^1 and s_1^2. Now substituting for:

$$\left(r_1, r_2, \ldots, r_{N_1},\ t_1, t_2, \ldots, t_{N_3},\ s_1^1,\ s_3, \ldots,\ s_{N_2}\right) \quad \text{and} \quad \left(r_1, r_2, \ldots, r_{N_1},\ t_1, t_2, \ldots, t_{N_3},\ s_1^2, s_3, \ldots,\ s_{N_2}\right)$$

into the quadratic equation 'q_2', we can set up:

$$F^1(s_2) = 0 \tag{2.A3}$$

$$F^2(s_2) = 0 \tag{2.A4}$$

where, F^1 and F^2 are quadratic expressions of s_2. Hence, in principle, we can find roots of s_2 as:

$$s_2^1,\ s_2^2,\ s_2^3,\ s_2^4$$

Thus, we have four possible sets of roots, as follows:

$$r_i, s_1^1, s_2^1, \ldots\ldots\ldots\ldots, s_{N_2}, t_i$$
$$r_i, s_1^1, s_2^2, \ldots\ldots\ldots\ldots, s_{N_2}, t_i$$

$r_i, s_1^2, s_2^3, \ldots\ldots\ldots, s_{N_2}, t_i$

$r_i, s_1^2, s_2^4, \ldots\ldots\ldots, s_{N_2}, t_i$

We can proceed in this manner until 2^{N_2} sets of roots are obtained.

Step B

Consider that $r_1, r_2, \ldots, r_{N_1}, s_1, s_2, \ldots, s_{N_2}, t_2, \ldots, t_{N_3}$ are known values. Then substituting these values into equation 'c_1' will give:

$$G(t_1) = 0 \tag{2.A5}$$

where $G(t_1)$ is a cubic expression of t_1. In principle, we have three possible roots from t_1, let these be denoted t_1^1, t_1^2 and t_1^3. Now substituting for –

$r_1, r_2, \ldots, r_{N_1}, s_1, s_2, \ldots, s_{N_2}, t_1^1, t_3, \ldots, t_{N_3}$;

$r_1, r_2, \ldots, r_{N_1}, s_1, s_2, \ldots, s_{N_2}, t_1^2, t_3, \ldots, t_{N_3}$;

$r_1, r_2, \ldots, r_{N_1}, s_1, s_2, \ldots, s_{N_2}, t_1^3, t_3, \ldots, t_{N_3}$

into the cubic expression 'c_2', we obtain:

$$\left.\begin{aligned} G^1(t_2) &= 0 \\ G^2(t_2) &= 0 \\ G^3(t_2) &= 0 \end{aligned}\right\} \tag{2.A6}$$

where G^1, G^2, G^3 are cubic expressions of t_2. In principle, we can then obtain nine roots of t_2.

Let these roots be denoted by: $t_2^1, t_2^2, t_2^3, \ldots, t_2^9$. Hence, we have nine possible sets of roots, and continuing along these lines we get, 3^{N_3} possible sets of roots.

Step C

Since step A can be carried out with each of the possible 3^{N_3} sets of roots '$t_1, t_2, \ldots, t_{N_3}$', we can see that a total of $(2^{N_2} \times 3^{N_3})$ sets of roots can be established.

Step D

For all the $(2^{N_2} \times 3^{N_3})$ sets of roots in step C '$s_1, s_2, \ldots, s_{N_2}$' and '$t_1, t_2, \ldots, t_{N_3}$', we can set up linear equations $l_1, l_2, \ldots, l_{N_1}$ for $r_1, r_2, \ldots, r_{N_1}$. Hence, we have unique values for $r_1, r_2, \ldots, r_{N_1}$ for each of the sets of '$s_1, s_2, \ldots, s_{N_2}$' and '$t_1, t_2, \ldots, t_{N_3}$'.

Hence, the total possible sets of $r_1, r_2, \ldots, r_{N_1}, s_1, s_2, \ldots, s_{N_2}, t_1, t_2, \ldots, t_{N_3}$ is given by:

$$N_{set} = 2^{N_2} \times 3^{N_3} \qquad \text{for} \qquad i = 1, 2, \ldots, N \tag{2.1}$$

3

A Comparison of a Conventional RANS and a Lattice Gas Dynamics Simulation – a Case Study in High-speed Rail Aerodynamics

A Gaylard

Synopsis

In recent years, external aerodynamics Lattice Gas Dynamics (LGD) simulations have become a viable alternative to conventional Reynolds Averaged Navier–Stokes (RANS) CFD. Potentially, they offer a more straightforward route to fully automated mesh generation and the routine acquisition of unsteady flow data. Before embracing the technique, it is important to understand how it compares with conventional CFD in terms of total set-up and simulation time, the computational resources required, and the results obtained. In addition, how CFD simulation can support an experimental programme and the degree of correlation that can be expected between these approaches are other vital dimensions in the effective application of CFD.

This chapter explores these issues through a case study in high-speed rail aerodynamics. The fourth framework-funded RAPIDE project has afforded the opportunity of simulating the flow around a German high-speed train, the ICE2, using commercial codes that employ both techniques. In addition, comparison will be made with measurements made in the wind-tunnel.

3.1 Introduction

Train designers, operators, and regulatory authorities need to understand the aerodynamic performance of their trains if they are to address key safety and performance issues, for example:

- reducing energy consumption;
- assessing the likelihood of train overturning by strong convective wind gusts; and
- establishing safe working practices to ensure the security of people and objects at the trackside.

Traditionally, this has been done using wind-tunnels and relatively small scale models. However, these are subject to well known limitations principally centred on lack of kinematic similarity (i.e. Reynolds Number effects) and the difficulties of adequate ground simulation for such long slender bodies (1). Therefore, CFD methods are attractive in that they offer the potential for aerodynamic simulation which overcomes these limitations. However, substantial barriers need to be overcome if CFD methods are to fully realise their potential, in particular:

- predictive accuracy;
- total set-up and simulation time; and
- computational resources required.

In recent years the emergence of commercially available Lattice Gas Dynamics (LGD) software suitable for application to vehicle aerodynamics, has promised significant progress in overcoming these barriers. This chapter seeks to compare the performance of a commercial LGD code (2) (PowerFLOW), with a conventional Reynolds Averaged Navier–Stokes code (STAR-CD) on a simulation of the aerodynamic performance of a high-speed rail vehicle and thus gauge the progress made in overcoming the barriers to the application of CFD for external aerodynamics applications.

This case study in high-speed rail aerodynamics has been developed from work undertaken for RAPIDE[1] (3), a collaborative European project funded under the European Commissions' Fourth Framework programme. The project is led by DB AG and includes SNCF, FS and AEA Technology-Rail, along with the vehicle manufacturer, ADTranz. As part of this extensive project MIRA were tasked with two complementary activities, below, which support the calculation of the flow around a complete train (4).

- CFD simulation of the air flow around an ICE2 driving trailer.
- Wind-tunnel experiments at both $1/8^{th}$ and $1/15^{th}$ scale to establish the aerodynamic performance of the ICE2 driving trailer.

The following sections will briefly review the details of the vehicle geometry, RANS and LGD models and then seek to compare the results with those obtained in the wind-tunnel experiments.

3.2 ICE2 model

The ICE2 model was based on CAD data supplied by the manufacturer. Fig. 3.1 shows a complete train running in Germany. The work discussed in this chapter focussed on the flows around the leading vehicle and a trailing carriage.

Before either construction of the wind-tunnel or CFD models some features of the real vehicle were simplified to facilitate a direct comparison between the results obtained from the CFD models and reduced scale wind-tunnel tests. The principal features simplified were:

- corrugated inter-car gap corridor connection surface; and
- bogies and running gear.

1 Railway Aerodynamics for Passing and Interaction with Dynamic Effects

Fig. 3.1 ICE2 train

Figure 3.2 shows the modelling simplification made to the inter-car gap corridor connection surface. The corrugations present on the real vehicle have been replaced with a smooth surface taken at an averaged vertical position through the corrugations. This simplification was made as the prime consideration, for this project was to have similarity between wind-tunnel and CFD models to provide a clear indication of the capabilities of the numerical approach.

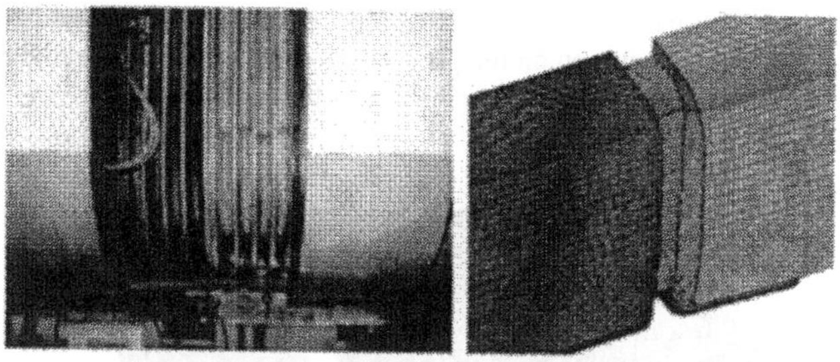

Fig. 3.2 Actual and modelled inter-car gap surfaces

In Fig. 3.3, the extent of the simplification of the bogies and running gear is revealed. These modifications reduced the complexity of the CFD calculations and also facilitated the construction of scale wind-tunnel models. This approach permitted geometric similarity between the CFD and wind-tunnel models to be, therefore, achieved economically.

3.3 CFD modelling details

3.3.1 Reynolds Averaged Navier–Stokes (STAR-CD)

The mesh representing the driving trailer and trailing carriage was constructed from mainly hexahedral elements with some 'wedge' cells created to capture a triangular (plan) 'spoiler' on the underside of the driving trailer nose. This was accomplished using the ICEMCFD/HEXA meshing tool.

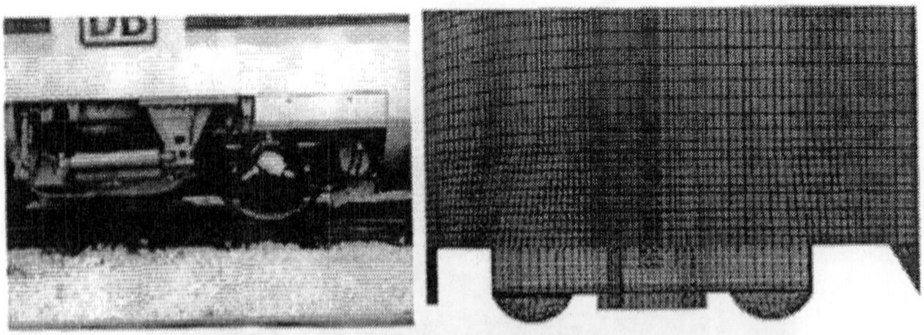

Fig. 3.3 Actual and modelled leading bogie

Centre-line symmetry was exploited to reduce the size of the calculation, a valid approach as the CFD simulations were all run with zero yaw onset flow. Regions of complex recirculating flow, such as the inter-car gap and the zone around the bogies, were further resolved with local embedded mesh refinement. Cell angularity was minimized using local and global 'O' grids that also facilitated a closely spaced mesh adjacent to the vehicle body enabling the maximum y^+ value on the vehicle body to be kept below 50, whilst respecting the minimum y^+ requirement of the wall treatment used (standard wall functions). The computational domain was sized to minimize numerical errors, the inlet boundary being placed eight vehicle heights upstream of the driving trailer nose and both the upper and side boundaries (constant pressure) set six vehicle heights from the train body. The 'floor plane' was simulated by using a sliding wall moving at freestream velocity (35 ms^{-1}). The final mesh comprised close to 2 400 000 fluid cells. The mesh in the vicinity of the train nose is illustrated by Fig. 3.4, this shows high levels of mesh resolution on the principal body radii and the 'wedge' cells used to capture the triangular (plan) spoiler on the underside of the nose.

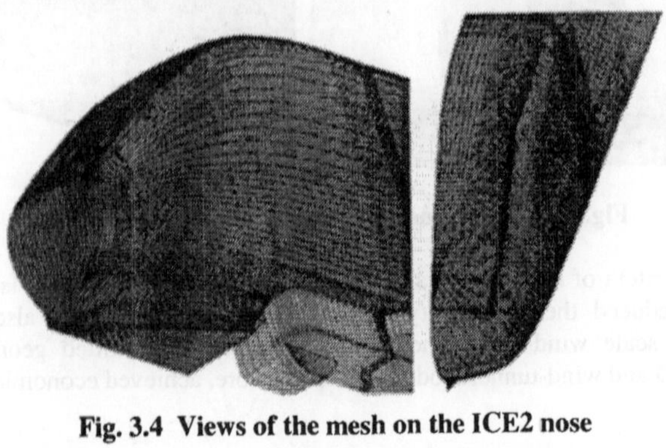

Fig. 3.4 Views of the mesh on the ICE2 nose

Turbulence closure was obtained via the RNG-derived κ–ε model. The third-order accurate MARS scheme was used to handle the spatial discretization of the convective fluxes. A $1/15^{th}$ scale CFD calculation was then run to convergence using STAR-HPC running on a 12 processor Silicon Graphics Origin 2000 compute-server.

3.3.2 Lattice gas dynamics (PowerFLOW)

A PowerFLOW model was set up by providing a triangular surface mesh, using ICEMCFD/TETRA. Once imported into the pre-processor, regions of additional refinement were specified, as can be seen in Fig. 3.5. This resulted in a lattice (mesh) comprising 16.46 million voxels (cells) and requiring 4 058 MB RAM.

Fig. 3.5 LGD model showing the region close to the train nose

Figure 3.6 shows the surfaces for the complete model. As there is no easy method for truncating the body using this solver, then two driving trailers with one intermediate carriage were modelled. This has the added benefit of providing a prediction for the wake flow.

Fig. 3.6 LGD model surfaces

The simulation was then run until body force coefficients became invariant with time, as illustrated in Fig. 3.7. Thus, the simulation 'converged' to a 'steady-state'.

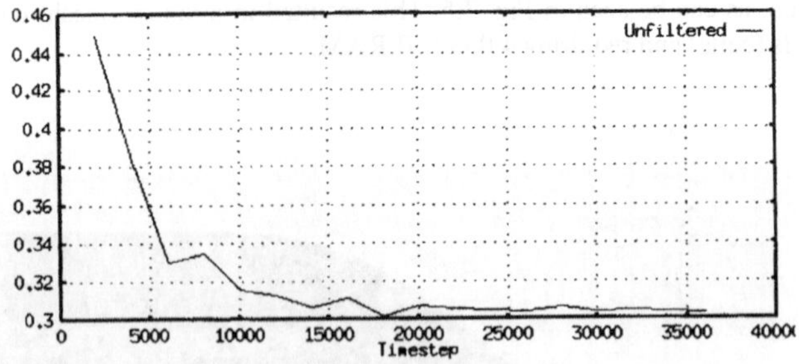

Fig. 3.7 Drag coefficient history for the LGD model

3.4 Wind-tunnel experiments

3.4.1 1/15th scale model

MIRA's Model Wind-Tunnel (MWT) was used in its moving ground configuration to undertake measurements of aerodynamic forces and surface pressures. This facility has the following general specification:

- 3/4 open jet test section, semi-open return (Eiffel).
- Jet nozzle of dimensions 2 m wide by 1 m high.
- Maximum test velocity 40 ms^{-1}.
- Average turbulence intensity in the empty test section of 1.1 per cent.
- Moving ground implemented using a 3 m long by 1 m wide belt with both belt suction and platen cooling.
- Boundary layer suction and tangential blowing upstream of the belt leading edge.
- Aerodynamic forces and moments measured using an external three-component balance (drag and one lift component are measured within the model).

Figure 3.8 shows the model mounted on an external aerodynamic balance via external supports. A trailing half carriage is also visible, which terminates in an ellipsoidal aerodynamic fairing. In this configuration, the nose of the model has been 'gritted' to promote early boundary layer transition, and both the driving trailer and half trailing carriage are connected to the balance for convenience during a flow visualization exercise. When force measurements were made the 'tail wire' was moved forwards, so only the driving trailer was mounted on the balance.

3.4.2 1/8th scale model

A further experimental programme was undertaken with a 1/8th scale model, utilizing MIRA's Full Scale (automotive) Wind-Tunnel (FSWT). The FSWT has the following specification.

- Closed test section, closed return (NPL type).
- Test section 7.94 m wide by 4.42 m high.
- Maximum test velocity 36 ms^{-1}.
- Average turbulence intensity in the empty test section 0.9 per cent.
- Fixed ground.
- Floor boundary layer control via a 'boundary layer fence' (or 'scoop').
- Aerodynamic forces and moments measured using a six-component underfloor balance.
- ±180 degree turntable.

The model was installed on the balance using a bespoke mounting frame, for which a drag tare correction was obtained. The general arrangement is illustrated in Fig. 3.9.

Fig. 3.8 1/15th scale ICE2 model in MIRA's MWT

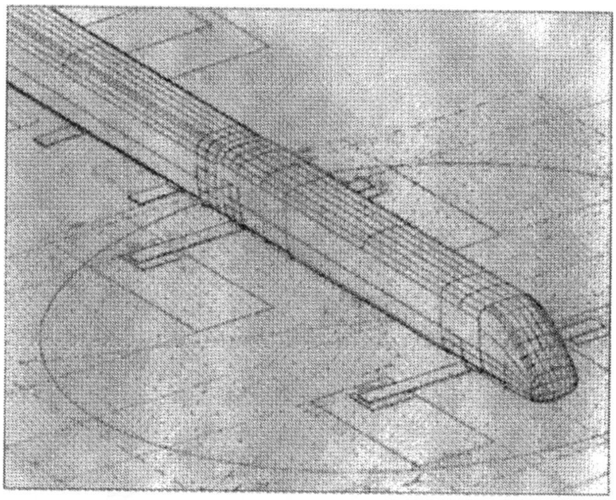

Fig. 3.9 CAD drawing showing model mounting arrangement

3.5 Comparisons

3.5.1 Set-up and run times

Before either the CFD models could be set-up or the wind-tunnel tests undertaken, a significant effort was required to produce representative CAD from the data provided by the manufacturer. The effort required to set-up and solve the model using the two CFD techniques is quantified in Table 3.1.

Table 3.1 Set-up and run times for RANS and LGD codes

CFD Method	Mesh and set-up model /Man hours	Run Model /Hours (on 10 processors)
RANS (STAR-CD)	105	68.47
LGD (PowerFLOW)	11.5	223.28 (includes 14.11hrs for discretization)

3.5.2 Surface pressure

Both of the scale models had static pressure tapings mounted along the centre-line of the driving trailer nose, running from the tip of the nose over the roof, ending directly above the downstream edge of the bogie cut-out. This enabled the collection of surface pressure data, which is presented in Fig. 3.10, along with the comparable data from both of the CFD models.

It is clear from Fig 3.10, that the most significant difference is between the two wind-tunnel experiments, rather than the CFD simulations and experiment. This arises, in the main, from two sources: Reynolds Number effects and wind-tunnel variability. The 1/15th scale model was tested at a Reynolds Number (Re_H) of 567×10^3, whilst the larger 1/8th scale model permitted a Re_H of up to 1.1×10^6 to be obtained during the test programme. Measurements of drag coefficient made on the 1/15th scale model with varying flow velocity indicated that even using remedial measures to promote transition, such as trip wires and grit, significant Reynolds Number sensitivity still existed. In addition, the variation that exists between different wind-tunnels, even when the same model is tested, is both significant and well known (**5**).

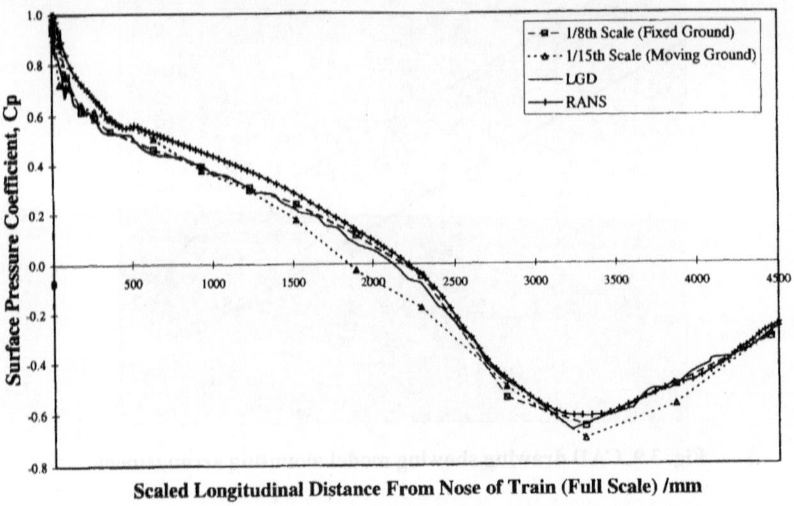

Fig. 3.10 Comparison of surface pressure coefficients (C_P)

Comparing the two CFD methods it can be seen that they are in close agreement over the suction zone associated with the leading roof radius, but the RANS solver over-predicts the surface pressure on the raked front face of the driving trailer, relative to the LGD solver. The evidence provided by both wind-tunnel experiments, particularly the measurements made on the 1/8th scale model, suggests that the LGD models results are more realistic in this specific regard. It is acknowledged, however, that since the LGD simulation incorporated the rear of the train whilst the RANS model did not, a fortuitous cancellation of errors resulting from the presence of this feature cannot be excluded in principle. In practice, the aspect ratio of the vehicle is sufficient that such an effect should not prove capable of causing the discrepancy between the measurements observed.

3.5.3 Flow structure

3.5.3.1 The importance of CFD as a flow visualization tool

CFD is, perhaps, at its most useful when providing comprehensive flow visualization. In many industrial applications it is more commonly applied as a flow visualization tool than a source of absolute quantitative data. Applied in this manner, CFD is often able to explain why fluid systems behave in the way they are observed to. Therefore, it is important to ascertain whether a relatively new tool, such as LGD solvers, are able to reliably elucidate flow structures. Two important regions of the vehicle are now examined to investigate the capability of the LGD solver to provide realistic flow visualization data. These regions are:

- inter-car gap; and
- tail.

3.5.3.2 Inter-car gap (ICG)

The cavity formed by two adjacent vehicles and the corridor connection between them can contain highly three-dimensional recirculating flow. This can, depending on the design of the vehicles, incur a significant drag penalty by transporting low-energy air from the underfloor and injecting it into the roof (or bodyside) boundary layer (1). Therefore, the ability to visualize the flow in this zone and predict any net mass transfer from the vehicle underside is an important test of the utility of a CFD code.

Fig. 3.11 shows the results of a simple flow visualization exercise using the 1/15th scale model in the moving ground wind-tunnel. The large tufts show the bulk flow passing over the ICG cavity driving a secondary flow in the ICG. The cotton tufts on the side of the corridor connection show a net downflow. Those on the roof reveal a reversed flow.

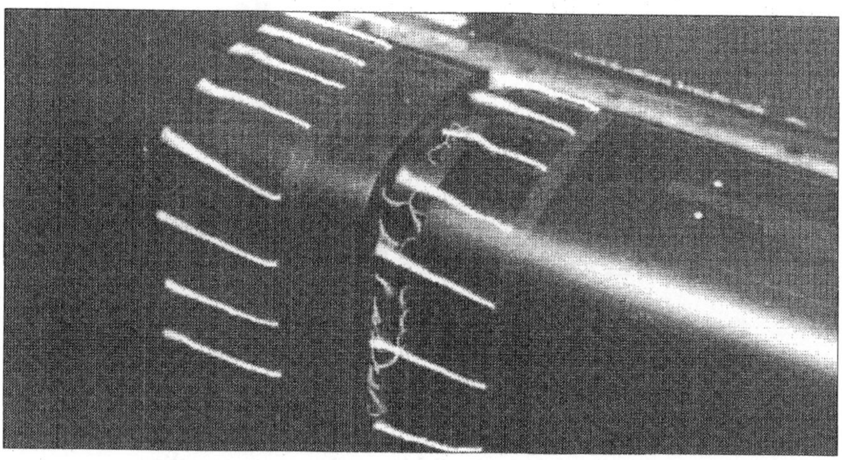

Fig. 3.11 Wool tufts showing flow direction in the ICG, 1/15th scale model

Both of these flow structures can be seen in Fig. 3.12, which uses flow 'streamlines' to reveal the three-dimensional flow pattern predicted by the LGD model. These can be seen spiralling down the ICG from top to bottom, as well as revealing reversed flow over the corridor connection roof.

Fig. 3.12 LGD model showing flow direction in the ICG

Figure 3.13 presents the predictions of the RANS model for flow direction in the ICG. In Fig. 3.13(a), the velocity vector plot shows flow reversal over the corridor connection roof with the formation of what is effectively a driven cavity flow. In Fig. 3.13(b), contours of vertical velocity are used to reveal the net downflow in the ICG, the lighter the colour on the plot the larger the downflow velocity.

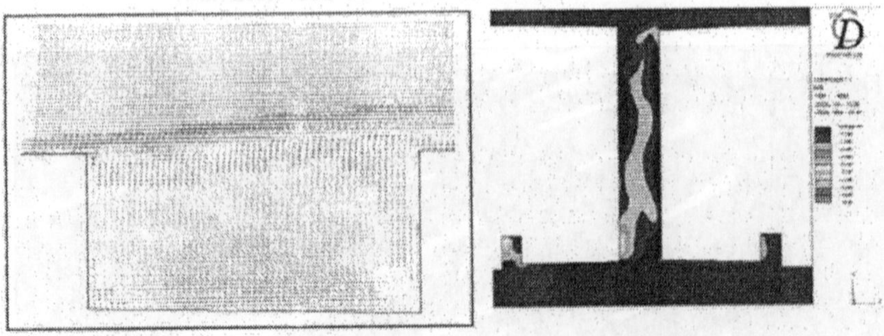

(a) (b)

Fig. 3.13 RANS model showing flow direction in the ICG using:
(a) Velocity vectors on the centre-line, and
(b) Contours of the vertical velocity component (in a plane mid-way between the centre-line and bodyside) with lighter colours representing downwards flow

Therefore, it is clear that both CFD techniques are able to visualize the flow structure in this region. However, it is worth noting that the LGD simulation is unsteady whereas the RANS solution has been run as a steady-state simulation. Thus the added CPU cost (Table 3.1) of the LGD method is justified, at least in part, by the opportunity to examine any flow unsteadiness that may be present.

3.5.3.3 Flow close to the tail of the train ('near-wake')

As stated at the outset, the primary focus for MIRA's involvement in RAPIDE and thus the case study developed from the project, was to model flow around the leading vehicle and first carriage of the train. However, it was more convenient to use the LGD model to represent a consist including a rear-facing driving trailer (i.e. tail car). Therefore, the data set from the LGD CFD exercise is more comprehensive than that obtained using the RANS code. This provides an added opportunity to assess the visualization capability of the LGD method by assessing how well it predicts the flow structure around the tail of the train. This is an important feature of the flow, as this is the early part of the vehicle wake. The wake of a train has a significant impact, along with the slipstream in general, on the security of people (both maintenance workers and passengers waiting at platforms) and objects at the trackside. Thus, from an engineering perspective the ability to determine the flow structure in this 'near-wake' region is an important one.

At present, as the RAPIDE project is not yet complete[2] data with which to correlate the LGD predictions are sparse, limited to the items listed below.

- Surface flow visualization on the $1/8^{th}$ scale wind-tunnel model rotated through 180 degrees with the ellipsoidal fairing facing into the flow (hence using the flow over the tail of a short train to represent that over a longer train).
- Limited full-scale flow visualization carried out by one of the other RAPIDE partners (**6**).

The LGD models prediction for the near-wake structure is shown in Fig. 3.14. This revealed a flow dominated by a pair of strong and concentrated vortices sitting along the trailing bodyside radii. These entrain flow around the bodyside radii and particularly from the rear face of the driving trailer. This latter effect results in the flow on the rear face of the driving trailer remaining attached until the tip of the tail is reached. Once shed from the driving trailer surfaces, the vortices then impinge on the (moving) ground.

A comparison between surface flow visualization performed using the $1/8^{th}$ scale wind-tunnel experiment and the LGD model are shown in both Figs 3.15 and 3.16. Figure 3.15(a) reveals the strong entrainment of fluid on the rear facing surface resulting in both attached flow and high angularity. These features are present in the LGD model, as seen in Fig. 3.15(b), but the CFD model predicts a lower degree of surface flow angularity, suggesting that the vortex structure is more tightly bound to the surface in the wind-tunnel experiment than is predicted by the LGD model.

2 In due course a comprehensive data set comprising full scale measurements, 1/15th scale wind-tunnel measurements over moving ground and a transient RANS simulation will be available (**3**).

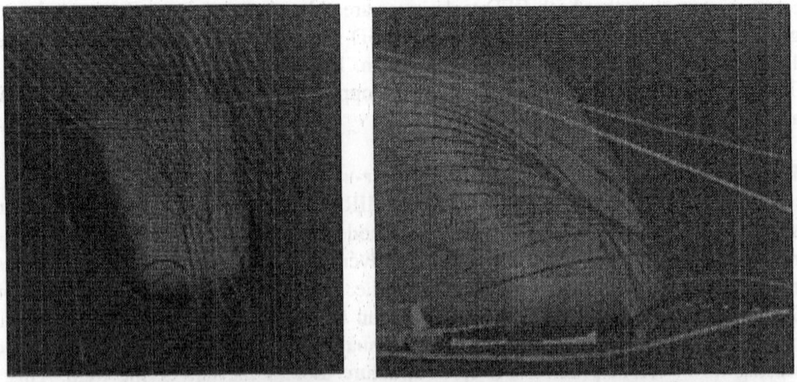

Fig. 3.14 LGD prediction of vortex structure in the near-wake

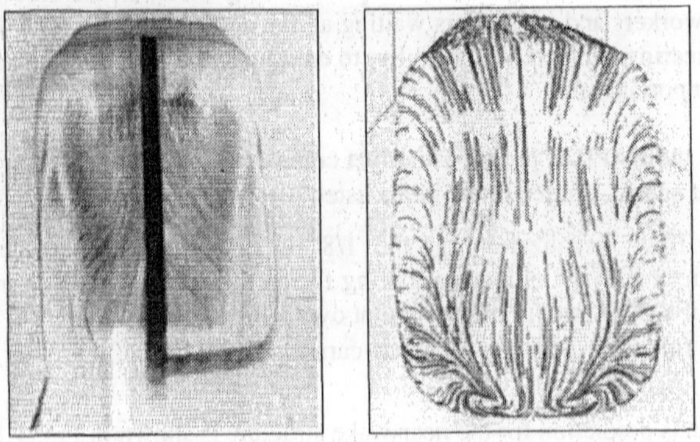

(a) (b)

Fig. 3.15 Surface flow visualization, (a), and LGD model, (b)

A view from the side is presented in Fig. 3.16, again indicating that the LGD model is predicting a similar flow structure, but with reduced flow angularity.

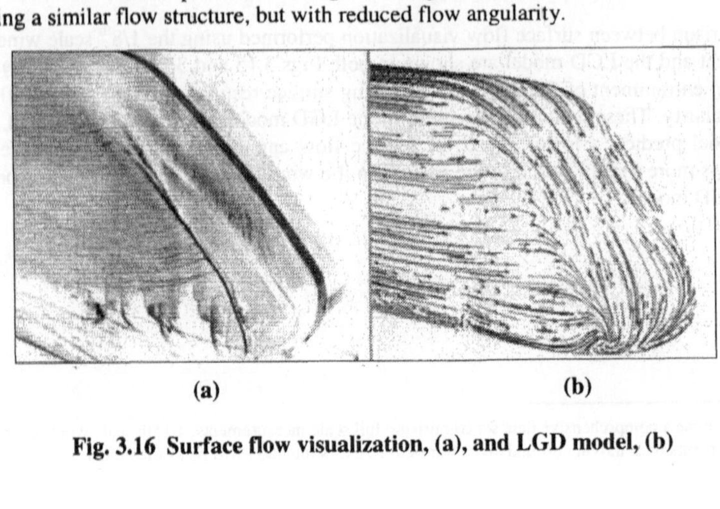

(a) (b)

Fig. 3.16 Surface flow visualization, (a), and LGD model, (b)

A final observation is that the position of the bodyside separation line is very similar in both LGD model and wind-tunnel experiment. The full scale flow visualization data presented in Fig. 3.17 confirms that both the 1/8th scale wind-tunnel experiment and LGD model correctly place the bodyside separation line. As more data becomes available it will be possible to determine whether the CFD model or 1/8th scale wind-tunnel experiment provides the best match with reality.

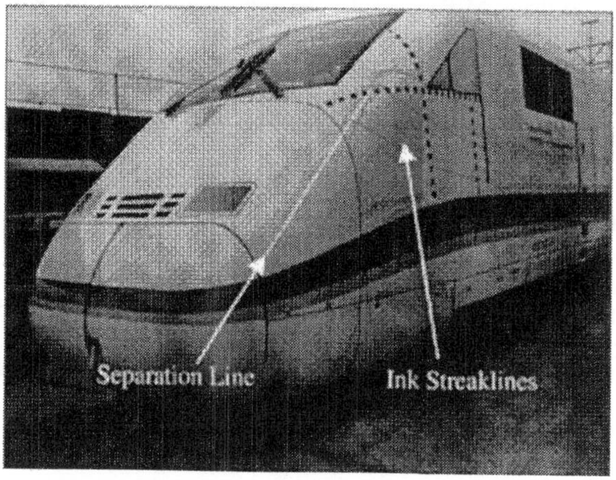

Fig. 3.17 Full scale flow visualization (6)

3.6 Conclusions

The following conclusions can be drawn from the foregoing:

- The rapid set-up time for the LGD code is a distinct advantage, particularly as one of the major roles for CFD is as a preliminary assessment tool prior to wind-tunnel testing.
- There is a large difference between the level of computational resources required by both CFD codes. The LGD model captured three times the geometry modelled using the RANS solver but required approximately four times the RAM. Similarly, the CPU time required was greater by a factor of 3.3. However Fig. 3.7 implies that the calculation had reached a steady state at around 25 000 time steps. Thus this additional CPU requirement could have been reduced by around 29 per cent. Hence, deploying the LGD methodology requires more RAM than an equivalent RANS model but the CPU requirement is not, in this case, significantly larger once differences in model size and criteria for termination of the calculation are accounted for.
- The LGD model produced a better prediction of centre-line surface pressure distribution. Both CFD models gave more reliable results than the 1/15th scale model, due to Reynolds Number effects.
- Both CFD approaches correctly predicted the flow direction in the inter-car gap.
- The LGD code appears to have produced a reasonable prediction for the near-wake structure.
- Within the context of the RAPIDE project, both CFD codes produced predictions on a timescale which delivered the data before the wind-tunnel work could be undertaken. Thus, decisions about the design of both wind-tunnel and full scale experiments could

be guided by the CFD predictions. If CFD is to be a viable tool in the armoury of the ground vehicle aerodynamicist then this is where CFD modelling must fit in the project lifecycle.

3.7 Acknowledgements

The author wishes to thank MIRA and the steering committee of the RAPIDE project for permission to disseminate this information. Particular thanks are due to G Matschke (DB AG), for both the original CAD and the full scale flow visualization experiment, and W H Lee (MIRA) for constructing the surface and volume meshes. Finally, M A Brown (MIRA) and R Gregoiré (SNCF) are both thanked for their careful proof reading and suggestions.

3.8 References

(1) **Gaylard, A. P., Howlett, A. B.,** and **Harrison, D. J.** (1994) Assessing drag reduction measures for high-speed trains, *Vehicle Aerodynamics*, Loughborough University of Technology, UK, 18–19 July, The Royal Aeronautical Society, London.

(2) **Anagost, A., Alajbegovic, A., Chen, H., Hill, D., Teixeira, C.,** and **Molvig, K.** (1997) Digital physics analysis of the morel body in ground proximity, SAE Paper No. 9800371, *SAE International Congress and Exposition*, Detroit, Michigan, USA, 24–27 February.

(3) **Schulte-Werning, B., Matschke, G., Gregoiré, R.,** and **Johnson, T.** (1999) Rapide: a project of joint aerodynamics research of the European high-speed rail operators, ES1-3, World Congress on *Railway Research*, Tokyo, Japan, 19–23 October.

(4) **Matschke, G., Schulte-Werning, B., Fauchier, C., Gregoiré, R.,** and **Gaylard, A. P.** (1999) Numerical simulation of the flow around a six-coach high-speed train, ES1-2, World Congress on *Railway Research*, Tokyo, Japan, 19–23October.

(5) **Hucho, W.-H.** (Ed.) (1998) *Aerodynamics of Road Vehicles, from Fluid Mechanics to Vehicle Engineering*, Fourth Edition, Society of Automotive Engineers Inc., pp. 710–713.

(6) **Matschke, G.** Private Communication.

4

Mesh Generation – the Ricardo Philosophy

S M Sapsford and **M E A Bardsley**

Synopsis

Automatic hexahedral mesh generation with VECTIS has been around for almost ten years. For applications related to IC engines, challenges in terms of movement of the computational grid have been addressed even more effectively with the introduction of 'Boundary Mesh Motion'. This chapter presents the philosophy behind the methods adopted and the integration with CAD and other analytical methods to maximize the effectiveness of practical CFD. Examples showing how CFD has been integrated into the design/development cycle at Ricardo are shown.

4.1 Introduction

The main barrier preventing CFD becoming an everyday development tool has traditionally been the time it takes to generate computational grids. When considering the specific application to internal combustion (IC) engines, specific issues arise relating to mesh motion, including the piston and valve motion, small clearances (low valve lifts and valve/piston clearances), and re-entrant geometry. Numerous workers within the literature, e.g. (1), have constantly referred to target times for mesh generation and analysis to achieve acceptability within the design and development environment.

This chapter describes the industry-proven Ricardo mesh generation philosophy, the advanced mesh motion capabilities, typical applications, and elapsed times for mesh generation and analysis, and demonstrates that CFD can be very effectively integrated into design and development programmes.

4.2 VECTIS CFD software

VECTIS is a CFD code that has been developed specifically to address fluid flow problems in the vehicle and IC engine industries. Special attention has been given to treating complex geometries efficiently in order to:

- Have greater impact during design and development;
- Reduce timescales;
- Reduce cost.

The use of VECTIS is focused on the following applications:

- Coolant flows;
- In-cylinder flows;
- Flow/distribution in inlet systems (direct coupling to WAVE 1-D simulation);
- Flow/distribution in exhaust systems and catalysts (direct coupling to WAVE 1-D simulation);
- Thermal management (direct coupling to FLOWMASTER 1-D simulation).

Of these, it is geometrical complexity and moving meshes that pose the greatest challenge in terms of generating useful results within a reasonable timescale, and it is these applications that form the focus of this chapter.

4.3 Mesh generation

Mesh generation in VECTIS starts with the geometry of the flow domain defined by a triangulated surface model. The model is usually generated in a CAD system and is imported into VECTIS in one of two standard data exchange file formats: STL or VDA-FS. STL files contain a triangulated surface representation; VDA-FS files contain polynomial surfaces which are triangulated by VECTIS to a user-defined tolerance.

The triangulated surface has to form a completely closed solid, without holes, gaps, or overlaps. To ensure that this is achieved, VECTIS provides a number of tools for the manipulation and repair of triangulated surface models. The reason for this requirement is that the mesh generator relies heavily on testing to see whether a point lies inside or outside the model. If the model is not a closed solid, ambiguities can arise in these inside/outside tests.

The mesh structure employed by VECTIS is Cartesian with local mesh refinement, and truncation of boundary cells. The mesh density is primarily controlled by the 'global mesh' which is defined interactively by the user. Figure 4.1 shows the Cartesian global mesh around a cylinder. In general, the spacing of the mesh lines in each of the three co-ordinate directions can be non-uniform.

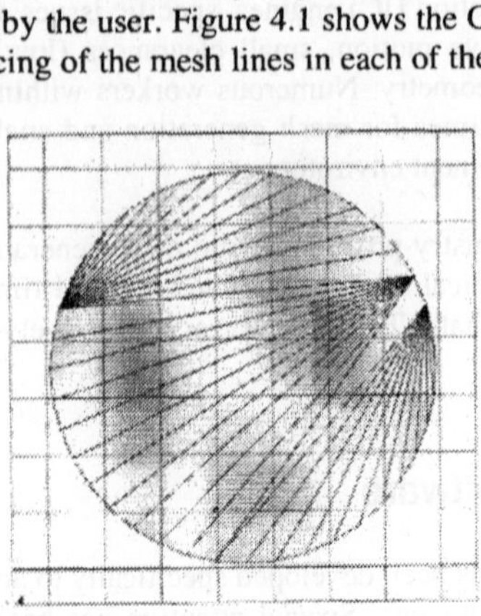

Fig. 4.1 Triangulated surface model with global mesh

Mesh generation proceeds in three stages. First, all global cells which lie entirely outside the model are eliminated (Fig. 4.2(a)). Then, the global cells are subdivided or 'refined' near the boundary in order to more closely approximate the shape of the flow domain (Fig. 4.2(b)). Finally, the refined cells are truncated to conform exactly to the original boundary (Fig. 4.2(c)).

Fig. 4.2(a)

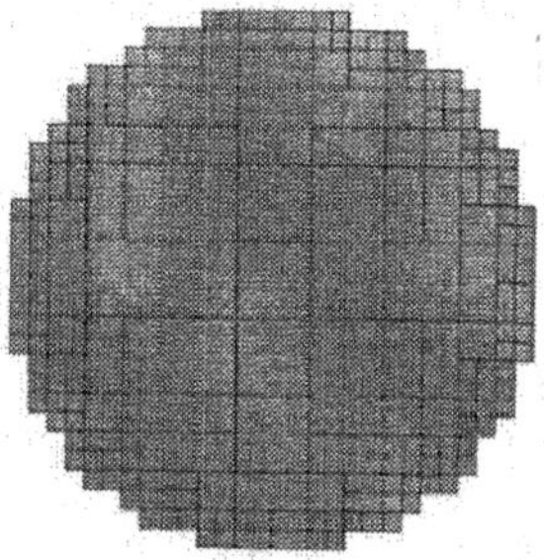

Fig. 4.2(b)

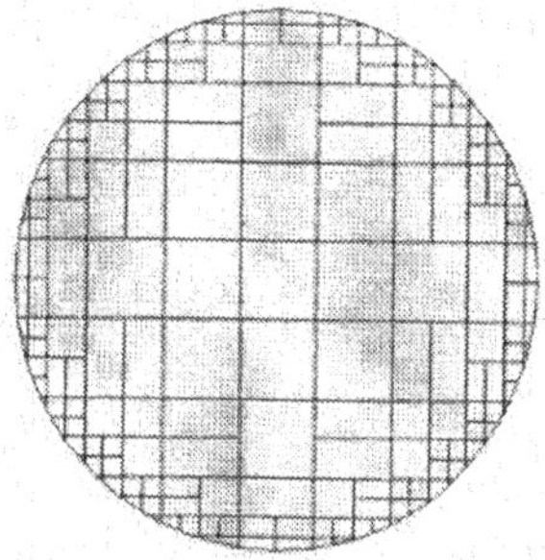

Fig. 4.2(c)

The amount of cell refinement near boundaries is controlled by the 'refinement level' parameter which is specified by the user. This parameter dictates how many times a global cell may be cut in two. A refinement level of 1 allows a global cell to be subdivided into a maximum of $2 \times 2 \times 2$ refined cells, level 2 allows $4 \times 4 \times 4$ subdivision, level 3 allows $8 \times 8 \times 8$ etc. The example in Fig. 4.2 uses level 2 refinement. Refinement of a cell is not applied uniformly in all three directions, e.g. a cell may be divided in the x direction but not the y or z directions. This makes more efficient use of cells than a method such as an octree approach, in which the only option for cell refinement would be $2 \times 2 \times 2$ subdivision to produce eight smaller cells. Figure 4.2 exhibits a number of the different mesh configurations which can be produced by level 2 refinement. This example is only two-dimensional; in three dimensions, there can obviously be a very large number of possible refinement patterns. The refinement algorithm works by evaluating all the possible combinations which respect the boundary shape, and selecting the possibility which produces the fewest cells. The algorithm also ensures a gradation of cell sizes towards the boundary, thus ensuring that the smallest cells are close to the boundary, where gradients in the solution usually require the finest resolution.

Cell refinement can also be introduced by defining regions of the global mesh which have refinement applied to them regardless of the presence of the boundary. This block refinement can be useful for increasing the mesh resolution around particular flow features, without increasing the global mesh density. The example in Fig. 4.3 shows layers of level 2 and level 1 refinement around a small cylinder.

The truncated cells at the boundaries are, in general, completely arbitrary polyhedra. Their volumes and face areas are calculated accurately so that the mesh is a true representation of the original geometry. Special measures are taken in the VECTIS solver in order to accommodate the resulting mesh structure. One-to-many connectivity, arbitrary cell shapes and variation of cell sizes are all treated so as to ensure accurate solution of the flow field near the boundary. Figures 4.4(a) and (b) show a typical surface model and computational mesh for an in-cylinder flow domain where the block refinement regions can be clearly seen.

This mesh generation method is reliable and easy to use. The use of mesh refinement and cell truncation allows the mesh to capture geometrical details over a range of scales without producing a mesh with too many cells. It is completely automatic and requires no input from the user other than specification of the global mesh and the refinement level. It runs at a speed of 5000–10 000 cells per minute for most applications on a typical Unix workstation, and thereby contributes significantly to the goal of reducing the elapsed time for a CFD analysis.

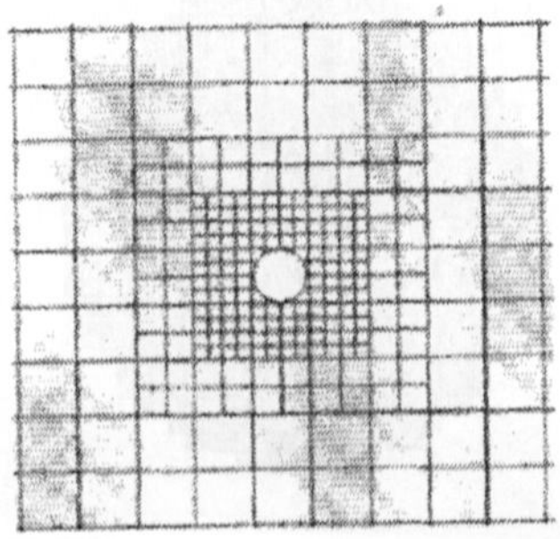

Fig. 4.3 Block refinement

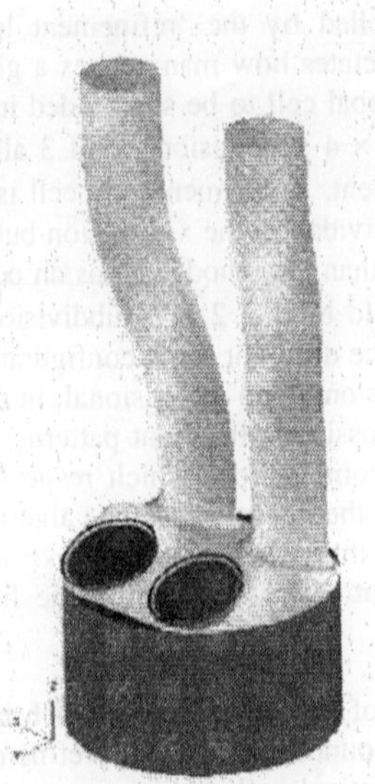

Fig. 4.4(a) Surface model

Fig. 4.4(b) Typical computational mesh

4.4 Geometrical complexity

A good example of extreme geometrical complexity is the underbonnet environment. Analysis of the underbonnet region is becoming a routine part of vehicle thermal management as the thermal performance of the vehicle becomes ever more demanding. Styling, packaging, performance, and comfort all place their own demands on a cooling system that is already challenged.

In the underbonnet region there are numerous individual components all of which must be captured in adequate detail. Figure 4.5 shows a slice through a typical vehicle.

Fig. 4.5 Underhood geometry

The powertrain, exhaust system, ancillaries, fan shroud, pipes etc. can all be seen (due to the way they are modelled the heat exchangers and fan are not visible in this figure). The majority of the time is actually spent preparing the geometry prior to mesh generation. This is required to repair errors, check for component interference and also to offset thin surfaces to give them realistic geometrical thickness. Once this is completed the mesh can be generated automatically in approximately four hours (~2 million cells). Figure 4.6 shows a slice through a typical mesh.

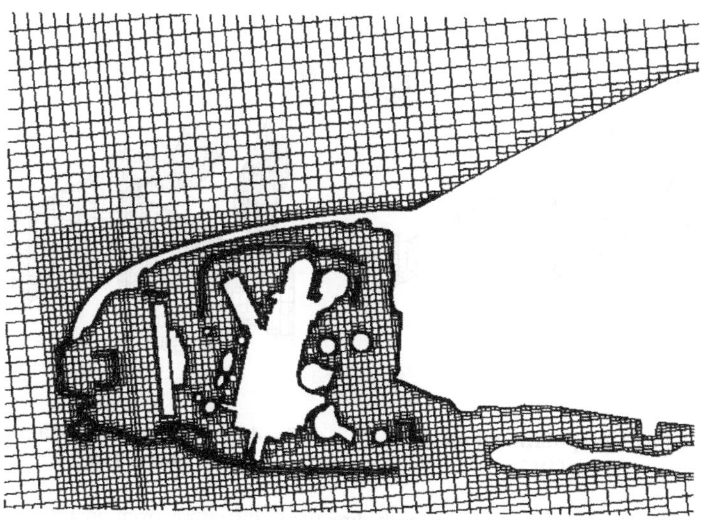

Fig. 4.6 Slice through underhood mesh

The other challenge for this type of analysis is the disparity in cell sizes between the extremities of the vehicle model in a wind tunnel compared to the detail required under the bonnet. For this, the mesh refinement approach described earlier is exploited. For the mesh shown in Fig. 4.6, the typical cell size in the underhood region is 20 mm with refinement down to 5 mm where extra resolution is required. Outside the underhood region where a section of a virtual wind tunnel is modelled, the cell sizes are graduated away from the underhood region up to around 1 m.

4.5 Mesh motion

Having created a computational mesh, the user promptly faces another, sometimes daunting challenge – getting it to move. The mesh motion in VECTIS is handled using the unique boundary mesh motion feature.

Using this technique, the nodes on the moving boundaries (valves and piston in this case) move with the correct displacement according to valve lift curves and piston displacement, the latter being calculated by VECTIS automatically from user-supplied stroke and con-rod length. Nodes on fixed boundaries such as the combustion chamber and valve guides are restrained and hence do not move. Nodes on the liner and valve stems act as an interpolated boundary between the moving and fixed boundaries.

As the boundaries move, the internal mesh structure automatically deforms in order to minimize the distortion of each individual cell (Figs 4.7(a) and (b)). This is performed by requiring that the displacement of any given point in the mesh is the mean of the displacements of surrounding points: mathematically, this means that the displacement's Laplacian is zero. This is exact in the limit of a dense mesh with large numbers of nearest neighbours. Discretization errors can arise because of the finite spacing of the mesh but these are resolved by iterative application of the mesh no-crossover condition to yield the final legal mesh. The calculation is automatically continuously monitored for excessive cell distortion (which could lead to mesh inversion) so that the calculation can be re-zoned onto a new mesh when required.

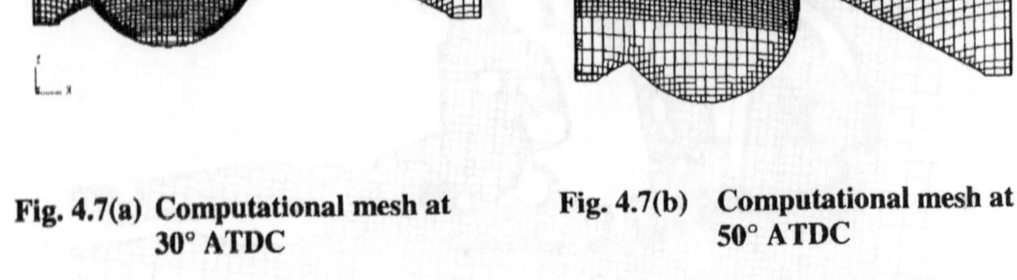

Fig. 4.7(a) Computational mesh at 30° ATDC **Fig. 4.7(b) Computational mesh at 50° ATDC**

In addition, meshes can be moved in both directions; distorting from an initially Cartesian grid and also progressing from a distorted mesh towards a Cartesian grid. The latter feature, known as reverse boundary mesh motion, is useful for the compression stroke, particularly inlet valve closure and piston approach to TDC, but a combination of reverse and forward motion is particularly useful for valve overlap.

4.6 Validation

Validation of the new boundary motion feature has been carried out on both gasoline direct injection (G-DI) engines and high-speed direct injection (HSDI) Diesel engines. The results shown below are for a G-DI engine are compared with PIV measurements from the Ricardo dynamic flow visualization rig (DFVR) (2) and LDA measurements obtained in an optically accessed engine as shown in Figs 4.8(a) to (c). The comparison is described in detail in (3).

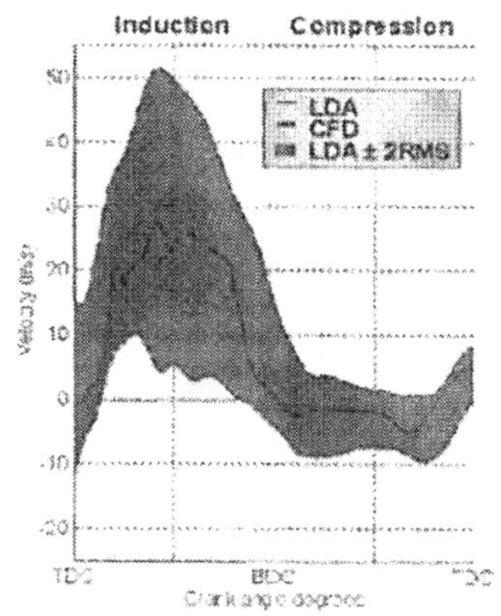

Fig. 4.8(a) Comparison of CFD and LDA results

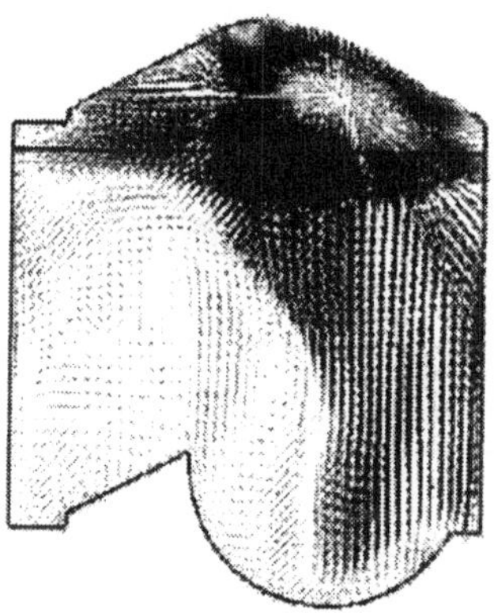

Fig. 4.8(b) CFD results at 120° ATDC (DFVR simulation)

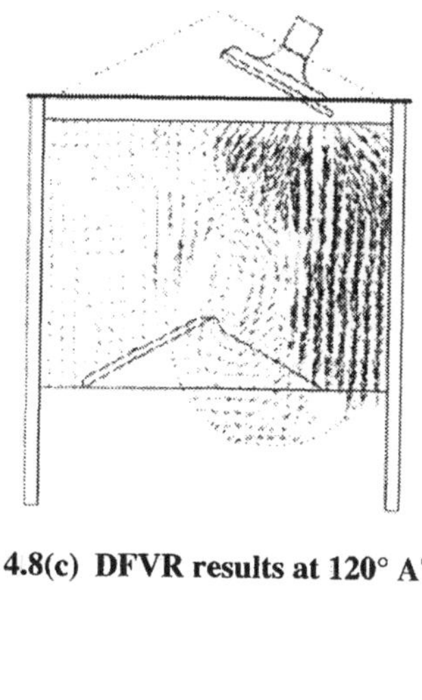

Fig. 4.8(c) DFVR results at 120° ATDC

Similar results have also been obtained for a swirling HSDI engine, again comparing with DFVR and LDA results (Figs 4.9(a) to (c)).

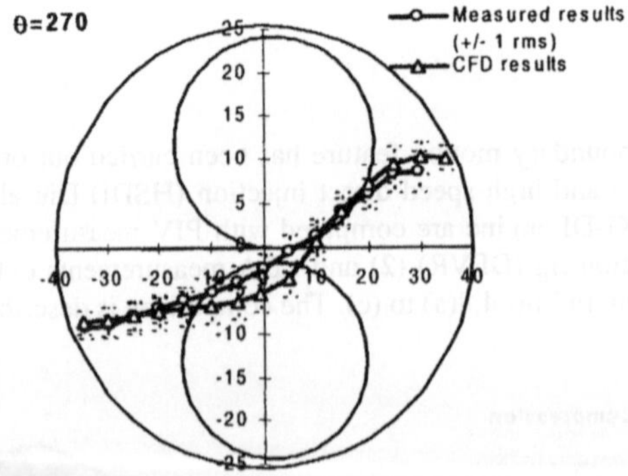

Fig. 4.9(a) Comparison of CFD and LDA swirl velocity results

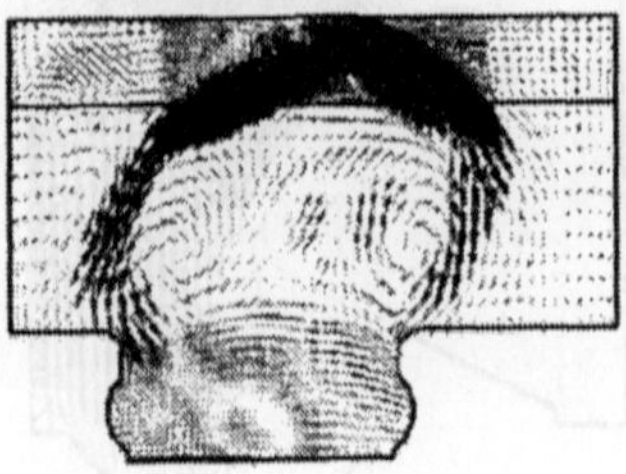

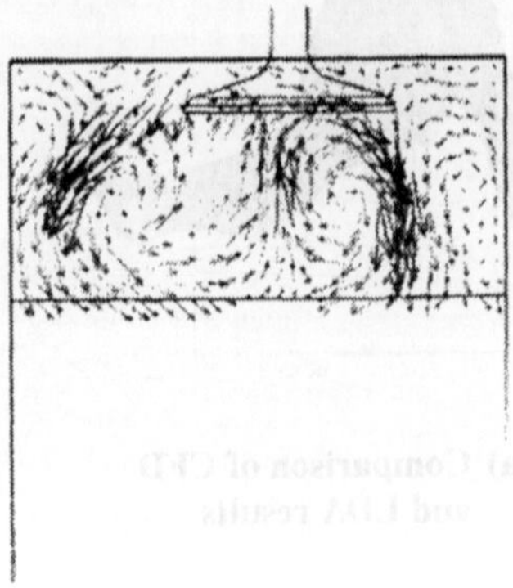

Fig. 4.9(b) CFD results at 70° ATDC (DFVR simulation)

Fig. 4.9(c) DFVR results at 70° ATDC

4.7 Typical application

The application to G-DI engines is of particular relevance as this represents a major opportunity for CFD to have direct influence on the development of a 'new' combustion system. The performance of a G-DI engine depends strongly on the interaction of a number of variables including:

- fuel injection characteristics;
- fuel spray targeting;
- injection timing;
- ignition timing/location;
- combustion chamber geometry;
- charge motion.

A typical programme will involve CFD analysis at the concept stage to determine successful strategies and then continued support through the development phases, investigating the effects of, for example, bowl geometry, injector characteristics, and operating conditions. This approach relies upon fast response. Set-up times including all mesh generation, mesh motion specification, and problem definition are typically less than two days, with an analysis or solution time of around one-and-a-half weeks (dependent upon complexity) for a complete intake and compression stroke including fuel spray. Of course, iterations on injector characteristics for stratified operation take considerably less time as only the last part of the compression stroke has to be re-analysed.

This approach has enabled all the major G-DI strategies to be analysed and has lead to VECTIS playing an important role in all major G-DI development programmes at Ricardo as shown in Fig. 4.10(a)–(d).

Fig. 10(a) Top entry (reverse tumble) with side injector (wall guided)

Fig. 10(b) Side port (swirl) with side injector (wall guided)

Fig. 10(c) Side port (forward tumble) with central injector (wall guided)

Fig. 10(d) Side port (forward tumble) with side injector (air guided)

4.8 Conclusions

Mesh generation is a critical factor determining the speed of all complex analyses. This chapter has described Ricardo's practical approach to mesh generation with particular emphasis on geometrical complexity and mesh motion.

This has enabled highly complex analyses to be carried out quickly and effectively, successfully integrating CFD into design and development programmes.

4.9 Acknowledgements

The authors would like to thank the Director of Ricardo Consulting Engineers Limited for their support and permission to publish this chapter. The authors would also like to acknowledge the help and support from their colleagues whose names do not appear in this chapter.

4.10 References

(1) **Bailly, O., Buchou, C., Floch, A. A.,** and **Sainsaulieu, L.** (1999) Simulation of the intake and compression strokes of a motored 4-valve S.I. engine with a finite element code, *Oil Gas Sci. Tech.* – Rev IFP, **54,** 161–168.

(2) **Jackson, N. S., Stokes, J., Heikal, M. R.,** and **Downie, J. H.** (1995) A dynamic flow visualisation rig for automotive combustion system development, SAE 950728.

(3) **Faure, M. A., Sadler, M., Oversby, K. K., Stokes, J., Begg, S. M., Pommier, L. S.,** and **Heikal, M. R.** (1998) Application of LDA and PIV techniques to the validation of a CFD model of a direct injection gasoline engine, SAE 982705.

(4) **Stevens, S. P., Bancroft, T. G.,** and **Sapsford, S. M.** (1999) Improving the effectiveness of underhood airflow prediction, Proceedings of the 1999 *Vehicle Thermal Management Systems Conference (VTMS4)*, C543/058/99, IMechE, London UK.

5

The Validation of Rapid CFD Modelling for Turbomachinery

H-H Tsuei, K Oliphant, and **D Japikse**

Synopsis

Good CFD calculations can be made to guide advanced turbomachinery design and development. The computing time and storage requirements, however, differ greatly from one computational approach to another and the resultant accuracy may well be debated. One specialist has suggested that most of the important effects in a turbomachinery blade row might be resolved using a coarse grid of only 30 000 nodes, while others insist on grids with ten times this node count. Arguments abound concerning the use of a wall law function as an engineering expedient. The present study draws on a set of seven (7) different stages, for which much measured data is available, and provides answers to these issues of sufficient depth to sensibly guide engineers in the economical and accurate utilization of their CFD tools. A base for rapid calculations is established; it is expected that the design future will focus intensely on agile, easy-to-use CFD as a base for advanced design development.

5.1 Introduction

Professor John Denton (**1**), observed that most of the important effects in a turbomachinery blade row can be resolved using CFD with a moderately coarse grid of 30 000 node points. This observation led to considerable thinking about and the eventual development of the pbCFD™ (Pushbutton CFD™ [1]) code now in use at Concepts ETI, Inc. (CETI). The code is built around the original Dawes (**2**), (**3**), solver (BTOB3D), which was introduced in the late 1980s as the first commercially viable CFD package for turbomachinery blade rows. Some 50 organizations around the world have come to use the Dawes code and, in most cases, rather extensively. Some companies to this day prefer this solver for bladed rows over any other CFD solver. After identifying at least five mechanisms by which the code could be accelerated by a factor of two, careful development work was undertaken to improve accuracy, accelerate the code, and make the flow code very easy, almost trivial, for engineers

1 Pushbutton CFD™ is a trademark of Concepts ETI, Inc.

to employ; hence pbCFD™ was created. Two computational errors were found in BTOB3D and corrected while creating pbCFD™; these corrections change the quantitative results.

A second interesting hypothesis concerning modern CFD modelling involves the method by which wall shear layers are resolved. Codes such as the Dawes BTOB3D or the TASCflow™ [2] code often use a logarithmic law near the wall to extrapolate the first grid point calculation down to the actual wall. While it is known that the log law forms an excellent representation of a two-dimensional boundary layer, preferably working outside of separation, it is also known that it is not a meaningful representation of fully three-dimensional (skewed) boundary layers. Nonetheless, it is commonly used and it is the general notion in the industry that if the first grid point is placed at a y+ in the range of 30–100, then very reasonable results are obtained. By contrast, other people feel that low Reynolds number turbulence models are preferable and allow one to compute the complete detail of the wall shear region. In this case, a y+ value on the order of one should be used to get numerical accuracy. The grid sizes for the latter may be quite large. Clearly, there is some conflict between the notion that Professor Dawes put forth (use the law of the wall) and the advocates for the low Reynolds number turbulence modelling, at least when viewed from the perspective of practical industrial calculations.

The present work covers a number of different test cases and compares some of the currently available computer codes. The modified Dawes code in the form of pbCFD™ and the FINE™/Turbo [3] code from NUMECA International, which utilizes the low Reynolds number turbulence modelling approach, are used herein. The ideas presented above require examination against data. It has become clear to the investigators that any single test case could be very misleading when looked at alone. Consequently, it was felt that a more statistical approach was necessary and that a number of relevant tests must be conducted. The first collection of data comprises seven different centrifugal compressor and centrifugal pump examples. These are simply the first group that was easy to assemble. It is a future objective to expand this set of comparisons up to approximately 20 different stages, hopefully before the end of 2000. In most instances, good data are available but work is required.

A fundamental rule of the work reported herein is to prohibit any parametric tweaking while using pbCFD™. In other words, once a basic set of modelling parameters is chosen, they must be used for the *whole set* of comparisons. Certain supporting studies have been conducted about sensitivity to the grid size and also to appropriate y+ values in order to provide useful background research. Nonetheless, the *final* comparison values to be used to judge the success of the pbCFD™ are based on a single set of operating parameters. In other words, no final tweaking of grids is to be permitted, no manipulation of the turbulence model shall be pursued, no messing with artificial viscosity is allowed, and, of course, a common approach to handling y+ near the wall is used for all cases.

A few items should be noted. *All of the work presented here must be considered preliminary at this time and is subject to further revision.* It is probable that some errors will be found both in data and in CFD which must be fixed. Indeed, for the PR-1.8 case we have repeated the traverse data three times in order to get data of sufficient accuracy that little error is being contributed from the laboratory; similar steps may be required for other cases. Likewise, the clearance flow or cavity leakage flow has not been modelled for pbCFD™. Further checks will be made and revisions reported at later times. These checks will include detailed matching of the actual distribution of traverse data at the impeller exit including total

2 TASCflow™ is a trademark of AEA Technology plc.
3 FINE™/Turbo is a trademark of NUMECA International, s.a.

pressure, static pressure, and yaw angle. Finally, it must be noted that the design reports referred to here are generally proprietary and are mentioned in the report for historical documentation purposes. All proprietary information has been eliminated from this report. Consortium stages such as the PR-4.5 and PR-1.8 are available to the participants of those consortia only. The Eckardt data is generally available throughout the world. The turbopump data are not available to the public.

It should also be noted that the calculations conducted here with pbCFD™ are converged to the design mass flow rate. An extension was made to the original Dawes BTOB3D program so that convergence to a desired back-pressure was replaced by convergence to a desired design flow rate. This increases the computational time modestly, while providing considerable design utility.

Any effort to validate CFD is still extremely complicated and one must be careful and not read too much into initial results. We intend to continue to refine and expand this work and to continue to question every detail that could influence results. The present work is focused specifically on pbCFD™ based on the historic Dawes code. It has not yet been possible to make calculations with FINE™/Turbo at the same level of intensity (i.e., rapid turn around). Consequently, any observation concerning FINE™/Turbo is on an early, preliminary basis. Nonetheless, the present FINE™/Turbo results were prepared by a thoroughly trained expert with this code and in some cases directly by NUMECA.

5.2 The PR45 compressor family

5.2.1 Background

The PR45 compressor family was studied through the 'Advanced Diffuser Consortium' project. Measured data, as well as computational results, were recorded in detail (**4**), (**5**). The design mass flow was 0.363 kg/s, with a rotational speed of 93 620 r/min. The inlet total pressure is 101.3 KPa while the inlet total temperature is 293 K. The inlet swirl and pitch angles are both zero. The geometry of this compressor is shown in Fig. 5.1.

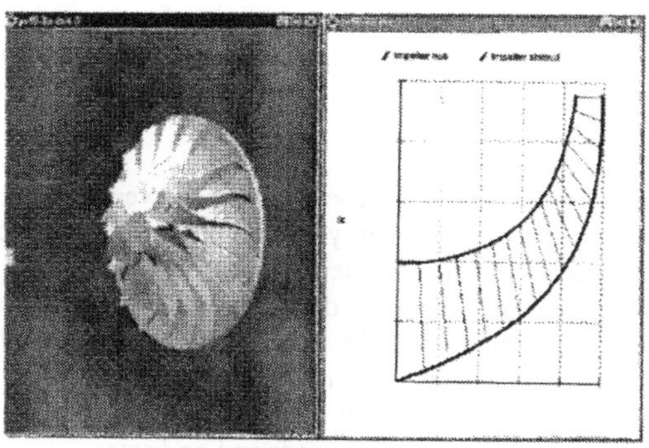

Fig. 5.1 The PR45 impeller

This particular example was of considerable interest because early calculations showed no separation in the passage, but as experimental results became available, it appeared that separation must be involved and some backflow or recirculation was likely. A series of studies was initiated which eventually led to the realization of some important modelling parameters and some errors in the BTOB3D code which needed correction. One of the early discoveries was that the full inlet duct must be realistically included with the impeller in order to obtain reasonable results. In the initial calculations, used when the design was first prepared, the inlet duct was only one-third the length of the actual duct used for test work. Subsequently, when the full inlet duct length was employed, separation was found in the impeller passages. As changes were made to upgrade the Dawes code into the pbCFD™ algorithm, more sensitivity studies for this particular configuration were conducted.

Detailed examination with the upgraded pbCFD™ of the computed flow field showed relatively large separation regions near both the splitter and main blade suction surface. The separation covered a depth from about the mean section to the shroud line. The revised pbCFD™ grid, with the extended inlet section, provided a greater boundary layer loss for the flow near the shroud region before entering the blade passages to result in the recirculation regions. Interestingly enough, with these separation regions present, the computed pressure ratio and efficiency were still much higher than measured data. This review raised a question about how accurate the original Dawes BTOB3D computational results were. Normally, we anticipate a much lower impeller efficiency if sizeable separation regions are present in the blade passages. It was observed that the original Dawes code neglects the energy diffusion (heat conduction and dissipation) terms in the energy equation, which could contribute to an observed overprediction in pressure and lead to a high efficiency.

To determine the effects of the energy diffusion terms on the calculation results, these terms were implemented into the Dawes solver. The energy diffusion terms are:

$$= \frac{\partial}{\partial xi}\left[(k+k_t)\frac{\partial}{\partial x_i}T\right]$$
$$+u_i \bullet \left[(\mu+\mu_t)\left(\frac{\partial u_i}{\partial x_j}+\frac{\partial u_j}{\partial x_i}-\frac{2}{3}\frac{\partial u_i}{\partial x_j}\delta ij\right)\right]$$

Key results are displayed in Table 5.1. The 100 per cent speed line data for rotor efficiency was used to compare with the computational results. All data for the first three (of four) cases in this study utilize full traverses (p_0, p, and α) just downstream of the impeller in a vaneless diffuser (the fourth case utilized total pressure probes in the throat of a subsequent vaned diffuser).

As can be seen in this table, the computed efficiency is on the average of two points above the measured values with the inclusion of the energy diffusion terms (compared to a much higher efficiency prediction obtained without these terms). An additional error affecting viscous evaluation of splitter blades was also discovered and fixed. Although the implementation of the energy diffusion terms and the splitter fix improved the results of this case, this also means the modified Dawes solver results *will* bring in a new perspective, which will impact on experienced Dawes code users. More study may be needed to look at the effects of the energy diffusion terms with a range of specific speeds and grid sizes.

5.2.2 y+ and grid sensitivity study

To better understand the effects of $y+ = y\sqrt{\tau_w/\rho}/(\upsilon)$ and grid sensitivity on the original Dawes solver solution quality, a study was undertaken. For detailed description of y+ and turbulent boundary layer physics, please refer to reference (**6**). Different values of y+ were obtained by either varying the grid stretching function while the total number of grid points was fixed, or by increasing the grid number in the hub-to-shroud and blade-to-blade directions. The grid node number in the meridional direction remained unchanged because y+ variation depends on the first grid spacing to a wetted surface in the hub-to-shroud and the blade-to-blade directions. The Dawes solver uses the algebraic Baldwin–Lomax turbulence model coupled with a wall function for turbulent flow simulations. Such an approach requires that the first grid point be located in the log layer region in order for the wall function to provide a reasonable wall shear stress calculation.

Table 5.1 Measured efficiency compared with pbCFD™ predictions

Mass Flow Rate, kg/s	Measured Rotor Efficiency	Computed Rotor Efficiency with Energy Diffusion Terms (pbCFD™)	Computed Rotor Efficiency w/o Energy Diffusion Terms
0.388	0.867	0.878	0.905
0.385	0.867	0.890	0.914
0.365	0.867	0.895	0.923

The first approach to obtain various y+ values is to use different grid stretching factors in the hub-to-shroud and blade-to-blade directions. The pbCFD™ default grid size was used. Table 5.2 summarizes three different stretching factors and their corresponding computational results. The y+ values in this table are average numbers throughout the blade passage. A larger grid stretching factor means the grid will be clustered more heavily near a surface. All the computations were performed on a Pentium 400 MHz platform and converged to within 1 per cent of the design flow rate of 0.363 kg/s.

Table 5.2 The effects of y+ on the solutions, based on the default grid size of 21 × 71 × 21, but different stretching factors using BTOB3D

Stretch Factor	Mass Flow, kg/s	y+	η_{tt}	p_{02ma} *Pa*	p_{2ma} *Pa*	T_{02m} *K*	M_2	1 % err CPU	2 % err CPU
1.2	0.3662	87.33	0.8602	6.082E5	3.167E5	520.91	1.012	20	15
1.3	0.3640	51.47	0.8520	6.043E5	3.113E5	521.87	1.021	38	30
1.4	0.3629	30.57	0.8587	6.130E5	3.144E5	522.41	1.025	42	33
1.5	0.3658	18.21	0.8578	6.156E5	3.159E5	523.36	1.025	45	39

(Pressures are mass averaged)

The pbCFD™ default grid and stretch factor (1.2) gives a y+ close to 90. Knowing that the wall function was designed to apply in the log layer, ideally y+ < 100, it is appropriate to apply this y+ when using wall functions. However, this y+ value is more on the high end of the wall function application criteria. Increasing the stretch factor to 1.3, 1.4, and 1.5 would provide smaller y+ values of about 50, 30, and 20, respectively. The computed efficiency was within a few tenths of a point from one another, and the computed flow variables at the TE were in good agreement with each other. The measured rotor efficiency was about 0.867. The computed results are all in good agreement with the measured efficiency (recall that leakage is not yet included in this study). One noticeable difference is that the CPU time usage goes up when the y+ value decreases. The stronger the grid stretching is, the larger the cell aspect ratio is. This situation creates the so-called 'acoustic stiffness' condition (which means the signal propagates much faster in one direction than another) and makes it difficult for the solver to converge quickly. For pbCFD™ design screening, the default grid size (21×71×21) with the default stretching factor (1.2) is therefore, recommended and suffers no loss in accuracy.

The next step is to investigate the second possibility of reducing y+: increase the grid points in both the hub-to-shroud and the blade-to-blade directions, while keeping the stretching factor fixed at 1.2. The calculated results are shown in Table 5.3.

Table 5.3 The effects of y+ on solutions, based on the same stretching factor but different grid sizes using BTOB3D.

Grid Size	Mass Flow, kg/s	y+	η_{tt}	p_{02ma} *Pa*	p_{2ma} *Pa*	T_{02m} *K*	M_2	1 % err CPU	2 % err CPU
21×71×21	0.3662	87.33	0.8602	6.082E5	3.167E5	520.91	1.012	20	15
31×71×31	0.3648	28.36	0.8423	5.982E5	3.105E5	522.92	1.015	87	65
41×71×41	0.3628	8.04	0.8235	5.969E5	3.116E5	527.84	1.010	267	191

(Pressures are mass averaged)

In this table, as the grid increases, the y+ values decrease significantly, from about 90 to less than 10. A consistent trend was observed. For the medium grid size (31×71×31) case, the y+ is about 30 and provides a very close prediction to the measured rotor efficiency. For the fine grid case (41×71×41), with the fact that the y+ already falls in the viscous sublayer (y+ < 10), coupled with the large grid size, the wall function produces excessive viscous stress near the wall region to cause the loss to be higher than measured, hence resulting in a lower rotor efficiency. It is recommended to not allow a first y+ value to fall in or near the viscous sublayer when using the wall function. Although the medium grid size case gave a good prediction, the only trade-off is the CPU time requirement. The CPU time needed to converge the medium grid size case to the design flow was increased by a factor of four. This CPU time requirement goes up exponentially when grid size increases.

From Table 5.1 and other results of this study, it is observed that the rotor performance prediction does not depend on the y+ value alone. For example, a case using a 21×71×21 grid, with a stretching factor of 1.4 and a second case of 1.2 (31×71×31), with a stretching factor of 1.2 are representative cases with y+ about 30. The predicted rotor efficiency for these two cases was about one and a half points apart from one another. The latter predicted a lower total pressure and Mach number at the TE, while the total temperature was almost the same. This indicated that the finer grid produced a larger loss near the wall regions to result in this discrepancy. *We learn from Table 5.2 that the use of the pbCFD™ default grid provides a very reasonable first approach, while a medium grid size might provide a finer solution compared to test data.* Further increasing the grid density in an attempt to reduce y+ to within the viscous sublayer is not recommended when the wall function is used.

The tests were conducted using two different grid systems (31×71×31 and 41×71×41) with the stretching factors of 1.2, 1.3, and 1.4. This was done to further investigate the effects of grid size (moving in the direction of a numerical grid independence) versus the effects of wall y+ values. The biggest effect found, however, was the tendency for large grid systems to force the first grid point into the laminar sublayer or into the laminar to turbulent transition regime, therefore significantly changing the computed results. No conclusions were reached concerning grid independence; the problem concerning a y+ value in the range of 30 to 100 was reinforced. Additionally, the changing value of y+ throughout the computational iterations was examined. As a rule, the value of y+ can drop anywhere from 10 per cent to 50 per cent from the initial calculation to the final converged result as the computational process proceeds and hence, one must be careful, once again, to choose a grid system with a sufficiently high initial y+ (on the order of 70 to 100 is recommended, recognizing that the values will decrease during the iterations). A future development is clearly required: an automatic method must be included within the codes to continuously scan for y+ values and to be sure that the grid system is realistic in terms of the first y+ value adjacent to a wall. When such a procedure is available, then numerical grid insensitivity studies can be properly conducted.

5.2.3 FINE™/Turbo results

The FINE™/Turbo solver was also used to perform the calculation for the baseline case. Computational results of three different levels (coarse, medium, and fine) were included. The coarse mesh size was about 8 000 nodes, the medium mesh size was 52 000 nodes, while the fine mesh size was 382 000 nodes. The medium mesh size was comparable to the pbCFD™ default grid size. The mass flow error was within 1 per cent of the design flow rate for the fine mesh, about 1.5 per cent for the medium grid, and about 4 per cent for the coarse mesh. The computation for the medium grid needed 41 CPU minutes on a Pentium 200 MHz processor platform (equivalent to approximately 20 CPU minutes on a Pentium 400 MHz processor) to converge.

The FINE™/Turbo results also indicated that the separation regions existed near both the splitter blade suction surface and the main blade suction surface. These recirculation regions were also observed in BTOB3D and pbCFD™ solutions, as above. The FINE™/Turbo results (mass averaging is used for all CFD results) are summarized in Table 5.4. All of the efficiency and rotor exit pressures are low compared to the data.

Table 5.4 FINE™/Turbo prediction results for the PR45 baseline case

Grid Size	Mass Flow kg/s	y+	η_{tt}	p_{02ma} *Pa*	p_{2ma} *Pa*	T_{02m} *K*	M_2	Residual Convergence	1 % err CPU	2 % err CPU
8,000	0.3444	<5	0.6505	4.179	2.263	512.41	0.948	Less	N/A	N/A
52,000	0.3588	<2	0.7168	4.740	2.526	520.19	0.994	than 3 orders	N/A	20
382,000	0.3602	<1	0.6954	4.648	2.443	526.17	1.021		N/A	N/A

(Pressures are mass averaged)

5.2.4 Summary

Comparative results are as follows. Figure 5.2 illustrates the pbCFD™ computational grid and, while it definitely is not coarse in character, it is not a thoroughly resolved 'fine' grid. Figure 5.3 shows the streamlines along a surface that is very close to the main blade pressure side; it is located one computational surface or grid element away from the pressure side. The streamlines are very orderly and the flow is essentially collateral. Figure 5.4 displays a similar result but it is one computational station away from the splitter blade suction surface. In this case, a region of backflow or recirculation is observed in addition to the normal development of strong secondary flow. Similar results have been obtained with FINE™/Turbo as shown in Fig. 5.5. This computation, however, uses 382 000 node points, yet yields a qualitatively similar flow field.

The y+ and grid sensitivity showed that a very good pbCFD™ prediction could be obtained if the first y+ value is in the mid-range of the log layer, i.e., 70 < y+ < 100 (if the y+ is in the viscous sublayer, a combination of inappropriate modelling methodology would likely occur).

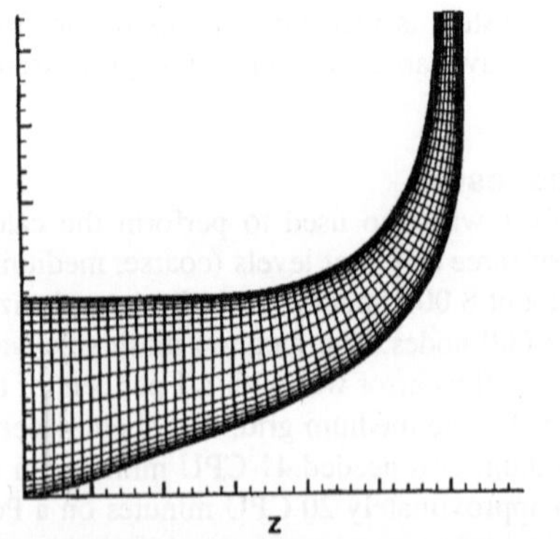

Fig. 5.2 pbCFD™ mesh for PR45 impeller, grid size = 21×71×21 = 31 311

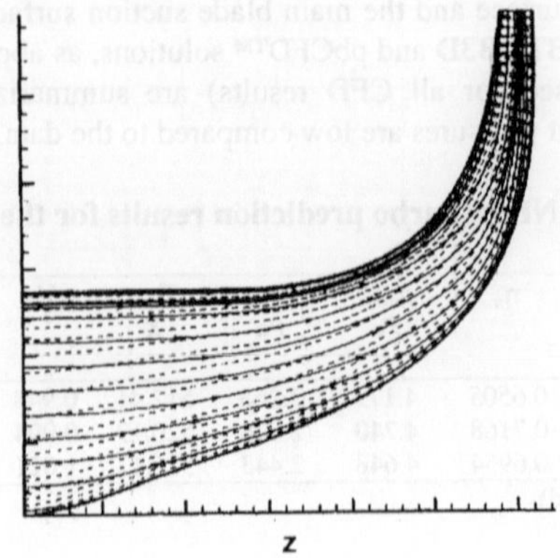

Fig. 5.3 PR45 impeller: pbCFD™ streamlines and velocity vectors at one computational surface away from the main blade pressure surface

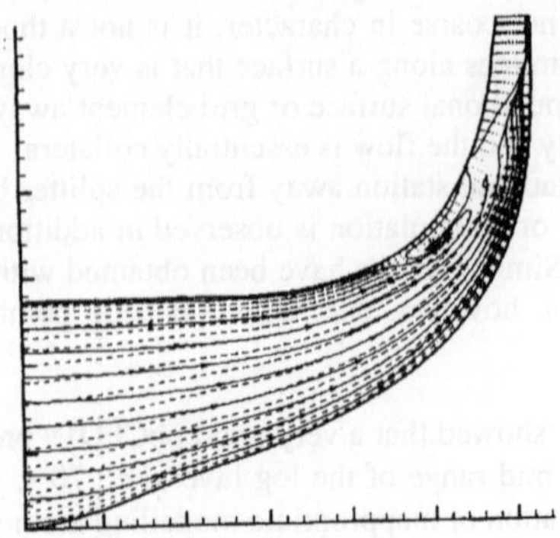

Fig. 5.4 PR45 Impeller: pbCFD™ streamlines and velocity vectors at one computational surface away from the splitter suction surface

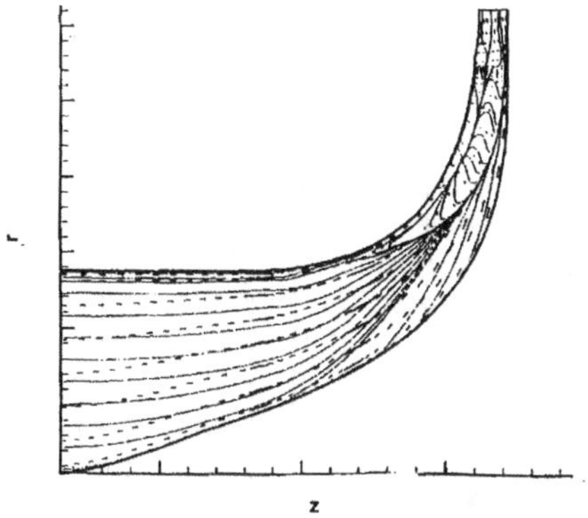

Fig. 5.5 PR45 Impeller: FINE™/Turbo results streamlines and velocity vectors at main blade suction surface, grid size = 382 000 nodes

A summary of rotor measured and predicted values is given in Table 5.5, the Dawes η is close; the FINE™/Turbo result (0.833) is a little bit low. The Dawes pressure ratio is high; the FINE™/Turbo pressure ratio is low. The FINE™/Turbo solver uses a low Reynolds number turbulence model to treat near wall turbulence; therefore, the first grid point must be placed in the viscous sublayer (y+ < 10) to provide a reasonable prediction. The coarse, medium and fine grids all have the first grid point in this regime. FINE™/Turbo predicted a total pressure ratio at the impeller exit of 4.6. This trend of underprediction was consistent for all three grid sizes, with the fine grid approaching a total of 400 000 nodes. The reason for this underprediction remains unclear. More effort to investigate the accuracy and quality of the FINE™/Turbo solution is needed.

Table 5.5 Summary

Mass Flow Rate kg/s	ηmeas	ηpbCFD™	ηFT	pr02m (meas)	pr02A (meas)	pr pbCFD™	prFT
0.363	0.867	0.895	0.833	5.12	5.59	5.98	4.59

(FT) – FINE™/Turbo (02m) – Impeller exit, mixed out state. (02A) – Impeller exit, mass averaged.

The pbCFD™ default grid using 32 000 nodes with a stretching factor of 1.2 (with y+ ~ 90) performed very well compared to the medium and fine grid size calculations. The rotor efficiency prediction, as well as major flow field characteristics, was captured by using this coarse grid. To begin examining the flow field of a new design, and also when running automatic optimization, it is highly recommended that the default grid size and stretch factor should be used to provide fast engineering solutions. The Dawes code was not designed to compute really large grid sizes. It works most efficiently when grid size is in the range of 30 K to 60 K nodes. The results of grid sizes over 100 K nodes could be misleading if the y+ becomes inappropriate. After course grid results, based on y+ information, one can fine-tune the analysis by using a moderately larger node count if necessary.

5.3 The PR18 compressor

5.3.1 pbCFD™ results

The low pressure ratio (pr = 1.8) compressor, PR18, was studied through two different consortia programs (4). The design mass flow was 0.15 kg/s, with a rotational speed of 43 560 r/min. The inlet total pressure and total temperature are 101.3 KPa and 300 K, respectively. There was no net swirl at the inlet. At the design point, the rotor efficiency was measured at 0.954 with a vaneless diffuser. The rotor pressure ratio was measured, by traverse with mass averaging, at 2.09. The geometry of the PR18 compressor is shown in Fig. 5.6.

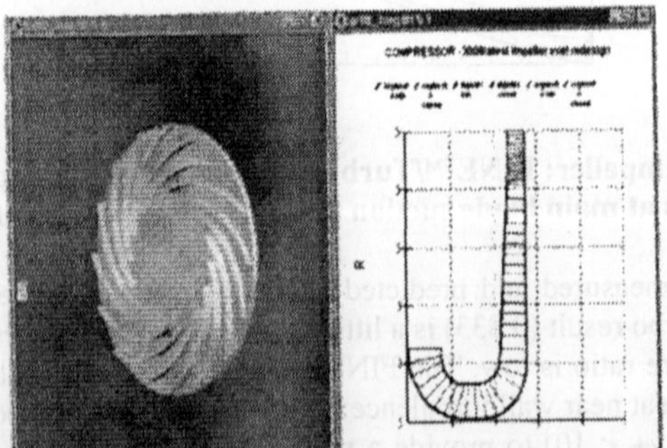

Fig. 5.6 PR18 stage

The pbCFD™ default grid size of 21×71×21 (total of 31 311 nodes) was used for this study. All CFD control parameters were set to default values, including the use of the energy diffusion terms. The PR18 compressor was composed of 16 main blades. The inlet extension duct is bent 180 degrees prior to the impeller leading edge, and the downstream element stretches to a location more than a full length of the impeller. The downstream radial extension region was cut short by using a CFD replacement segment to force more grid points in the impeller section to provide better predictions. It took 20 CPU minutes on a Pentium 400 MHz processor for the mass flow solver to converge the solution to within 1 per cent of the design flow, 15 CPU min. to within 2 per cent of the design flow. The final mass flow at the end of the computation was recorded at 0.1503 kg/s, almost exactly to the design flow of 0.15 kg/s. The predicted rotor efficiency was 0.933, about two points lower than the measured data. Front cavity leakage is not included. In the computed results, the flow was well-guided through the blade passages, with no separation region observed in the flow field. The computational results with or without the energy diffusion terms only showed two tenths of a point difference in predicted rotor efficiency, possibly due to the two-dimensionality of the rotor blade and the low pressure ratio. Table 5.6 summarized the mass-averaged pbCFD™ results.

An additional observation can be made from Table 5.6. Results are entered using only part of the inlet duct (90 degree inlet) and the full inlet duct which covers 180 degrees of turning. Once again, it is observed that modelling the full inlet is important. With only half of the inlet duct, the stage efficiency is computed at one point higher efficiency than with the full inlet duct.

Table 5.6 The pushbutton CFD predictions for the PR18 compressor

Inlet	Mass Flow kg/s	η_{tt}	p_{02m} Pa	p_2 Pa	T_{02m} K	M_2	Conv	1% err. CPU min.	2% err. CPU min.
90° inlet	0.1503	0.942 5	2.347E5	1.750E5	386.44	0.662	2.5	20	15
180° inlet	0.1520	0.933 1	2.355E5	1.776E5	386.58	0.648	2.0	30	20

(*FT*) – FINE™/Turbo (02*m*) – Impeller exit, mixed out state (02*A*) – Impeller exit, mass averaged

5.3.2 FINE™/Turbo results

The FINE™/Turbo CFD solver was also used to perform the calculation of the PR18 compressor. Table 5.7 shows the overall results of the different computations. The mesh size for one of the computations was comparable to the default mesh size in the pbCFD™ study and an additional finer mesh was also computed by NUMECA. As shown in Tables 5.7 and 5.8, the calculated efficiency varies from one to two points below the measured value. These NUMECA cases were run with only the 90 degree bend at the inlet.

Table 5.7 FINE™/Turbo prediction results for PR18 compressor

PR18	Mesh Size	y+	Mass Flow, kg/s	η_{tt}	p_{02m} Pa	Convergence
NUMECA	28 350	~8	0.15	0.92	2.27e5	> 3.0
NUMECA	212 954	~4	0.15	0.94	2.28e5	> 3.0

Figures 5.7 to 5.12 show the computed velocity vectors and streamlines. Figure 5.7 shows the standard computational grid for the pbCFD™ computation. Figures 5.8, 5.9, and 5.10 show the computed streamlines near the blade pressure surface (there are no splitters), in the mid-channel and near the blade suction surface, respectively. The core flow (Fig. 5.9) is quite orderly and reasonably collateral. The flow near the pressure surface and the suction surface shows the development of substantial secondary flow, with the distinct possibility of backflow near the end of the blade on the suction surface. It may also be observed that the inlet duct is designed with very good flow control and no evidence of separation is observed in the inlet, even though regions of moderate diffusion are unavoidable. This helps confirm a design hypothesis that such inlet ducts can be configured without inlet separations. Additionally, important results from FINE™/Turbo are shown in Figures 5.11 and 5.12. Very strong secondary flows are observed on the pressure surface with moderate distortion on the suction surface (Figs 5.11 and 5.12, respectively). Some very small regions of localized separation are observed. This separation at the inlet is not of concern in as much as it essentially represents a horseshoe vortex which must exist at the inlet regardless. This type of separation is not a classical two-dimensional or skewed boundary layer (3-dimensional) separation, but simply the development of an inevitable secondary flow. Near the impeller exit, a small separation zone appears to exist just at impeller exit, but with only a little backflow into the impeller exit. Further studies should be conducted by changing impeller tip depth, the diffuser pinch, and the possible use of splitters to change the blade number. In general, there is qualitative similarity between the results between the two different codes.

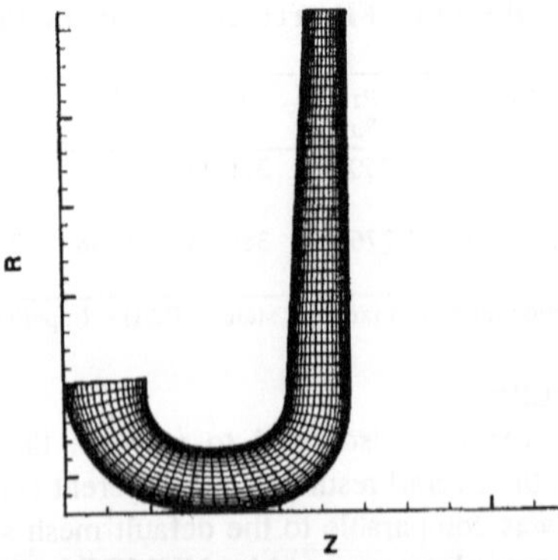

Figure 5.7 PR18 impeller with a 180° U-Bend inlet: pbCFD™ mesh, grid size = 21×71×21.

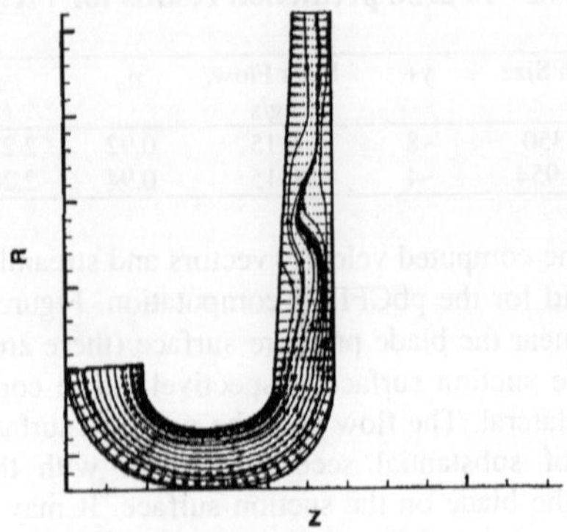

Fig. 5.8 PR18 impeller. pbCFD™ streamlines and velocity vectors at one computational surface away from main blade pressure surface.

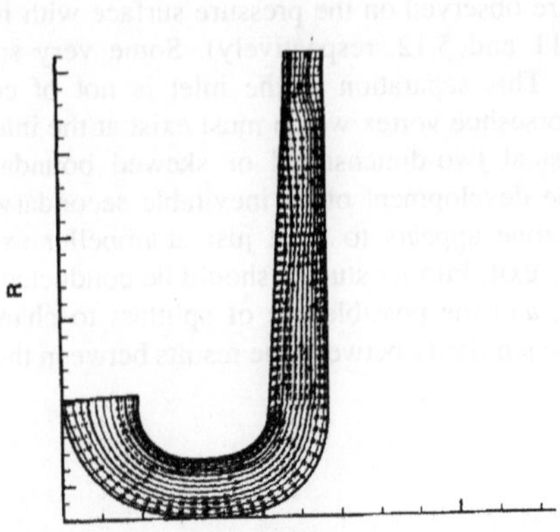

Fig. 5.9 PR18 impeller with a 180° U-bend inlet: pbCFD™ streamlines and velocity vectors at mid-channel.

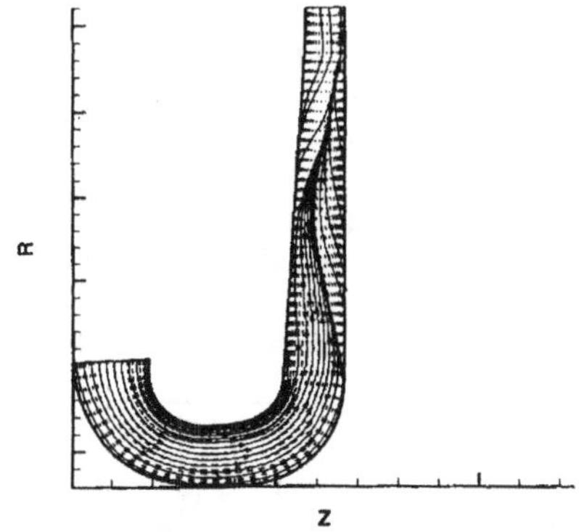

Fig. 5.10 PR18 impeller with a 180° U-bend inlet: pbCFD™ streamlines and velocity vectors at one surface away from the main blade suction surface.

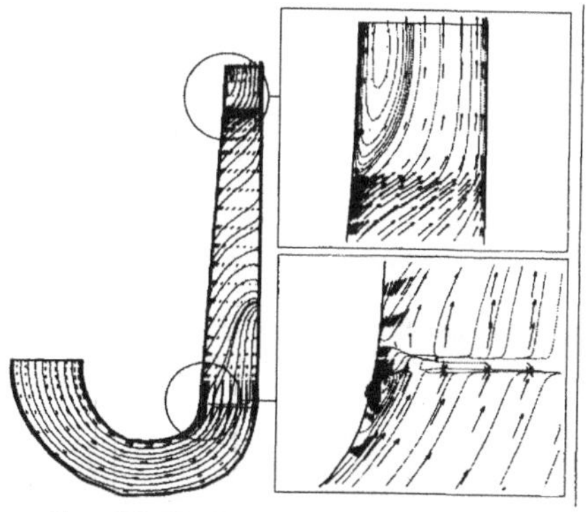

Fig. 5.11 PR18 impeller FINE™/Turbo CFD calculation, grid size = 42 650 nodes, blade pressure surface.

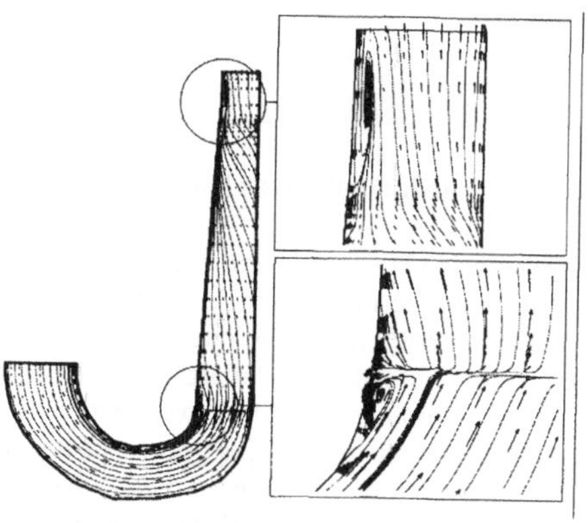

Fig. 5.12 PR18 impeller FINE™/Turbo CFD calculation, grid size = 42 650 nodes, blade suction surface.

5.3.3 Overview

Table 5.8 is a comparison between the pbCFD™ results and the FINE™/Turbo results at approximately the same mesh size. The CFD results over-predict the pressure ratio by about 10 per cent to 13 per cent. The FINE™/Turbo (run at NUMECA) is closer to the measured rotor efficiency than the FINE™/Turbo run at CETI which is about seven points too low. The results underpredicted the efficiency by about two points. Front cavity leakage must be modelled in future work but this will require a multi-block scheme.

Table 5.8 Summary of results from pbCFD™ and FINE™/Turbo

η_{meas}	$\eta_{pbCFD^{TM}}$	η_{numeca}	$pr_{02m,meas.}$	$pr_{02,\ pbCFD^{TM}}$	Pr_{2a}	$pr_{02,FTnumeca}$
0.954	0.933	0.935	2.09	2.32	2.15	2.24

meas – Measured; *FTnumeca* – FINE™/Turbo at NUMECA

5.4 Eckardt' radial compressor

5.4.1 Background

The so-called Eckardt compressor was an available stage that was carefully studied and reported by D. Eckardt (1987). The rotor has 0 degree backsweep. At the design point, the compressor operates at a rotational speed of 14 000 rpm with a design mass flow of 5.32 kg/s. The rotor is composed of 20 main blades, with no splitters. The inlet conditions were the standard operating conditions, with a total pressure of 101.3 KPa and a total temperature of 288 K. At 14 000 rpm, the measured rotor total pressure ratio was 2.180 and the stage efficiency was measured at 0.951. The geometry is shown in Fig. 5.13.

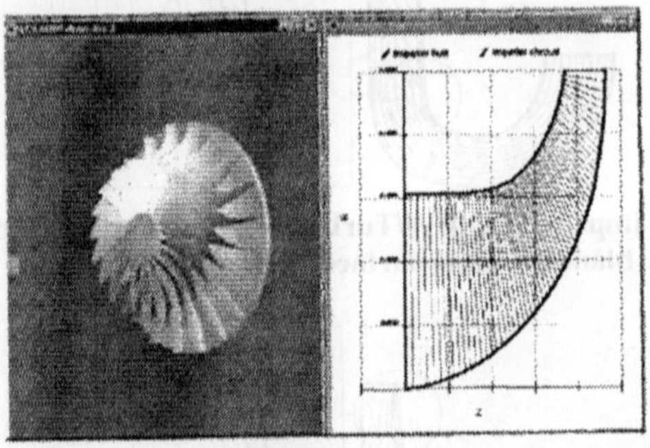

Fig. 5.13 Eckardt rotor '0' as used for CFD studies

The standard pbCFD™ default grid (21×71×21) and stretch factor (1.2) were used to perform the calculation. The energy diffusion terms were included in the computation. The calculation converged to within 1 per cent of the design flow rate and took about 40 CPU minutes on a Pentium 400 MHz platform. With a broader error margin, say 2 per cent, it still needed about 33 CPU minutes on the same computer. Detailed examination of the flow field solution indicated that the relative Mach number for a major portion of the compressor blade passage (from the pressure surface to mid-channel, from hub to mean area) fell below 0.2, a speed low enough to cause the unpreconditioned pbCFD™ code to converge slower. This was the reason that led to this uncharacteristically long CPU time. To show that the mass flow solver functions properly and that the CPU time is reasonable for this compressor calculation, a single fixed pressure ratio run was tested to investigate the convergence characteristics. At a fixed pressure ratio, the pbCFD™ code needed about 30 CPU minutes to converge the solution to this specific pressure ratio. The mass flow solver, which iterates the code to match a specified flow rate, converged the solution to within 1 per cent error of the design point in 60 CPU minutes. Because of the pbCFD™ code limitation, further improvement on CPU time at the design point for the Eckardt compressor is not likely to occur without the implementation of a preconditioning system to the pbCFD™ solvers (now in progress).

The computational results are tabulated in Table 5.9. The computation was conducted with the effects of the thermal diffusion terms included. The predicted rotor efficiency was 0.951. The measured total-to-static pressure ratio downstream of the TE was 1.471 at hub and 1.464 at shroud. The measured total-to-total pressure ratio 7 per cent (in radius) downstream of the TE was about 2.195. The computational data are in very good agreement with the measured data.

Table 5.9 The predicted Eckardt compressor rotor efficiency and flow properties at design point

	Mass Flow kg/s	η_{tt}	p_{02m} Pa	p_2 Pa	T_{02m} K	M_2	1% err. CPU min.	2% err. CPU min.
Computation	5.36	0.945	2.195E5	1.469E5	363.35	0.780	40	33
Data	5.32	0.951	2.205E5	1.447E5	363.5	—	—	—

5.4.2 y+ and grid sensitivity study for Eckardt compressor

The y+ and grid sensitivity effects on rotor performance prediction were also studied for the Eckardt compressor. Different values of y+ were obtained by either varying the grid stretching function while the total number of grid points is fixed, or by increasing the grid number in the hub-to-shroud and blade-to-blade directions.

The first approach to obtain various y+ values is to use different grid stretching factors in the hub-to-shroud and blade-to-blade directions. The default grid size was used. Table 5.10 summarizes three different stretching factors and their corresponding computational results. The y+ values in this table are averaged throughout the blade passage. A larger grid stretching factor means that the grid will be clustered more heavily toward a surface. All the computations were performed on a Pentium 400 MHz platform and converged to within 1 per cent error of the design flow rate of 5.32 kg/s.

In this table, when the stretching factor was 1.2, the resultant y+ was 175, a value much higher than the ideal wall function applicable range. When y+ is lowered to about 100 or less, but larger than 10, the predicted rotor efficiency, as well as other parameters, showed a very consistent trend. Although slight overprediction for the pressure ratio is observed, overall, the predicted results are in good agreement with the measured data. Indeed rotor efficiency prediction is now improved and the agreement is very good. Further stretching the grid into the viscous sublayer (y+ < 10) is not recommended.

Table 5.10 The effects of y+ on the solutions, based on the default grid size of 21×71×21, but different stretching factors

Stretch Factor	Mass Flow, kg/s	y+	η_{tt}	p_{02m} *Pa*	p_2 *Pa*	T_{02m} *K*	M_2	1 % err CPU	2 % err CPU
1.2	5.362	175.9	0.945	2.195E5	1.469E5	363.35	0.780	40	33
1.3	5.325	112.0	0.96	2.195E5	1.460E5	362.13	0.786	22	18
1.4	5.364	59.6	0.960	2.203E5	1.478E5	362.57	0.777	38	30
1.5	5.344	33.0	0.962	2.211E5	1.481E5	362.78	0.778	50	35
1.6	5.329	19.4	0.96	2.215E5	1.482E5	362.73	0.779	45	30
1.7	5.345	11.7	0.960	2.211E5	1.479E5	364.03	0.781	50	35

Table 5.11 The effects of y+ on solutions, based on the same stretching factor but different grid sizes

Grid Size	Mass Flow, kg/s	y+	η_{tt}	p_{02m} *Pa*	p_2 *Pa*	T_{02m} *K*	M_2	1 % err CPU	2 % err CPU
21×71×21	5.362	175.9	0.9455	2.195E5	1.469E5	363.35	0.780	40	33
31×71×31	5.364	60.41	0.9562	2.198E5	1.464E5	362.64	0.785	131	76
41×71×41	5.321	23.63	0.9621	2.218E5	1.456E5	361.88	0.790	305	191

In terms of grid sensitivity, a grid size of 31×71×31 was used. This selected grid size represents the total number of grid points doubled. Table 5.12 shows the computational results of the grid system of 31×71×31 with three different stretching factors. Consistent prediction results were observed when the y+ value was in the log layer. When y+ falls below 10, in the viscous sublayer, both the predicted total pressure and efficiency started to go up, a result of the wall function being cut off leading to insufficient loss near wall regions, as found earlier in the PR45 compressor study. The CPU time required to converge the solution to within 1 per cent of the design flow increases dramatically. The computational results showed good agreement with the coarse grid calculation shown in Table 5.9. It again proves that, for screening purposes, the coarse grid case (21×71×21), can be used sensibly in the design iteration and optimization process to make engineering design decisions.

Table 5.12 The effects of Y+ on the solutions, based on the default grid size of 31×71×31, but different stretching factors

Stretch Factor	Mass Flow, kg/s	y+	η_{tt}	p_{02m} *Pa*	p_2 *Pa*	T_{02m} *K*	M_2	1 % err CPU	2 % err CPU
1.2	5.364	60.41	0.9562	2.198E5	1.464E5	362.64	0.785	131	76
1.3	5.311	26.29	0.9664	2.212E5	1.459E5	361.99	0.794	191	81
1.4	5.296	7.55	0.9706	2.225E5	1.471E5	362.34	0.788	191	87

Figures 5.14 to 5.16 show the computed streamlines. Figure 5.14 displays the computational grid and, while it is a moderately coarse grid, it clearly holds a substantial amount of detail. Figure 5.15 shows the streamline distribution on the pressure surface and, as distinct from other examples in this study, the flow is not collateral along the pressure surface, but rather shows strong secondary flow development or boundary layer skewing. This passage has excess width with very low Mach numbers along the hub surface. Figure 5.16 shows similar streamlines along the suction surface and it is clear, with the strong skewing on both pressure and suction surfaces, that a strong secondary flow must be developed in the core of the passage.

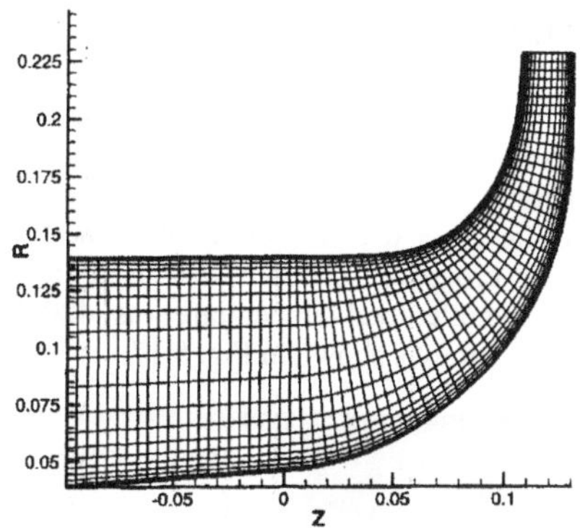

Fig. 5.14 Dawes mesh for Eckardt impeller: rotor '0' (0° backsweep), grid size = 21×71×21

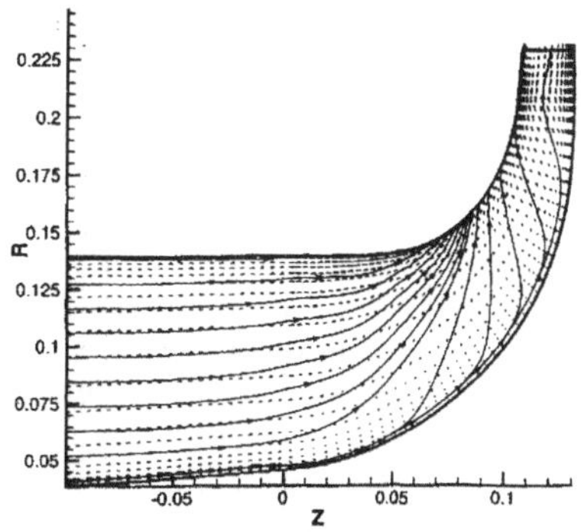

Fig. 5.15 Eckardt rotor '0': streamlines and velocity vectors at one computational surface away from the main blade pressure surface.

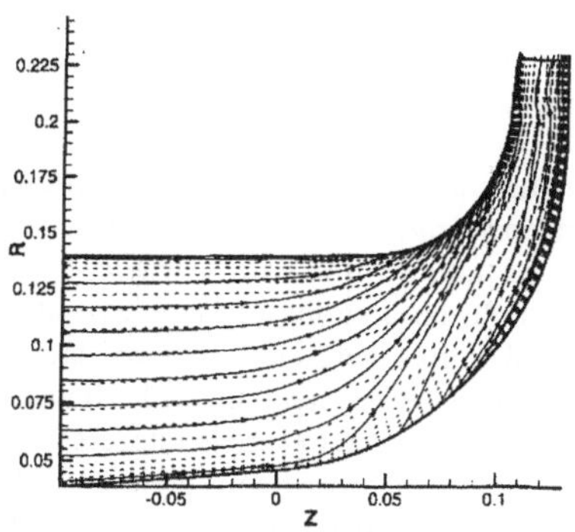

Fig. 5.16 Eckardt rotor '0': streamlines and velocity vectors at one computational surface away from the main blade suction surface.

5.5 Liquid hydrogen rocket turbopump

The liquid hydrogen alternative rocket turbopump for the Space Shuttle Main Engine (SSME) was developed at CETI for NASA Marshall Space Flight Center. The work was sponsored by a Phase I and II SBIR contract which is gratefully acknowledged. A matrix of geometry variations was considered in the rotor design process, with the final design being a bowed-blade turbopump impeller. A crossover diffuser was used in the stage. Because liquid hydrogen is a cryogenic fluid, the inlet total temperature was -420 degrees F. The geometry of the liquid hydrogen rocket turbopump is shown in Fig. 5.17.

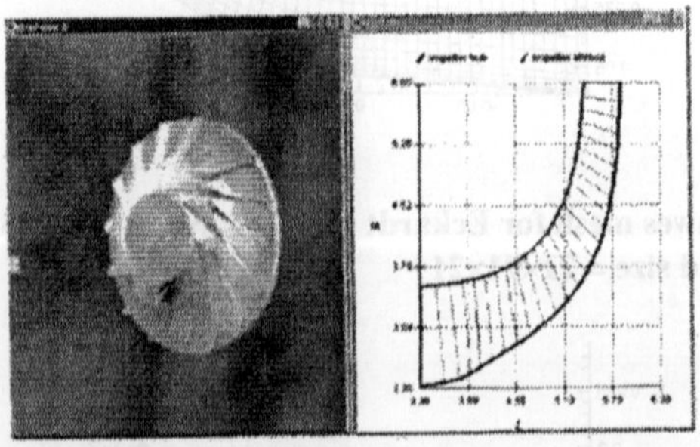

Fig. 5.17 Liquid hydrogen turbopump as evaluated with pbCFD™ (tested in water)

The CFD analysis for this project was conducted using both the pbCFD™ and the FINE™/Turbo solvers. For the pbCFD™ solver, a procedure was used to convert pump modelling to an air-equivalent compressor in order to perform the CFD calculations. This included matching all dimensionless groups including head, flow, and Reynolds numbers. In cryogenic liquid H_2, the Mach number is substantial and it was roughly matched. By using the rules of equal flow coefficient, head coefficient, and Reynolds number, fluid dynamic similarity between the pump flow field and the air-equivalent blower/pump was guaranteed. Therefore, the calculated streamlines and rotor efficiency, as well as flow angles, represent the flow field characteristics in the (liquid) pump blade passages. The computations used measured swirl angle (from appropriate prior rig tests) and flow angle as part of the inlet boundary conditions. The predicted rotor efficiency ranged from 0.96 to 0.97 using pbCFD™ and 0.976 to 0.996 using FINE™/Turbo for these design variations. The measured rotor efficiency was about 0.96. In the future, a proper preconditioning system will be added to pbCFD™ to support incompressible calculations.

The study used both solvers to evaluate a matrix (Table 5.13) of different design candidates. Two different design engineers independently rated the performance characteristics of each design as noted in the table. Using a rating from 1 to 5, where 1 would be an extremely benign or weak separation and 5 would be a strong or large separation, results of the two reviewers were averaged together and entries were made in each category, one for pbCFD™ and the other for FINE™/Turbo. Each reviewer came up with almost the same results (but their averaging yielded the x.5 values, as opposed to simple integers). Furthermore, each code generally led to approximately the same stage as being preferable. Using FINE™/Turbo, iterations, 3 and 4 looked the best and using pbCFD™ iterations, 4 and 5 looked the best. (The FINE™/Turbo efficiencies given in Table 5.13 were computed well after the design was completed and did not figure in the design choice. Early post-processing questions led to values that were not reliable. The current values seem a bit high and the last case seems unlikely.) It was concluded that Case 4 should be constructed. Case 4 was a very small modification of Case 3 with just a very slight change in the inlet shroud slope angle of ½ a degree. It is noted that the measured efficiency was approximately 96 per cent, as suggested in Table 5.14. It may be noted for completeness, that the calculations made with the pbCFD™ solver actually used a grid just over 50 000 node points; coarser grids using pbCFD™ are discussed below. FINE™/Turbo used 290 000 node points.

Table 5.13 Summary of BTOB3D and FINE™/Turbo CFD results for selected design iterations+

Case	Design Feature	BTOB3D Calculations	Separation Rating*	Impeller Efficiency From BTOB3D	FINE™/Turbo Calculations	Separation Rating*	Impeller Efficiency From FINE™/-Turbo
1	Nominal contours, blade angles, thickness.	Massive separation and reverse flow ahead of main blade and at splitter suction plus along main blade suction surface.	5.0	97.1%	Moderate separation on suction side of main blade near leading edge. Separation in front of main blade.	3.5	98.6
2	Inlet blade angle increased 5°.	Small upstream separation, moderate separation around L.E. of splitter pressure side, and new large separation on splitter and main blade suction side.	3.5	97.0%	Moderate separation around splitter and main blade L.E. *NO* main blade L.E. separation.	3.5	98.8
3	Contours refined, hub S-wall introduced, splitter moved forward.	Large separation on splitter suction surface; tiny backflow on main blade suction surface just before the splitter L.E. location.	4.5	96.7%	Probable small separation on suction side of splitter and maybe main blade near the shroud.	2.0	97.6
4**	Shroud inlet angle reduced 0.5°.	Tiny upstream separation, moderate splitter suction side separating strong skewing along shroud.	3.5	97.0%	Modest separation on suction side of main blade and slight separation on suction side of splitter blade near the shroud.	2.5	98.2
5	Radical changes in β distribution. Reduced turning in inlet, modest shroud ramp. Reduced shroud blade angle by 2.5°.	Moderate separation in front of main blade leading edge, moderate separation in front of and along splitter suction side and strong separation along main blade suction surface.	3.5	96.9%	Modest separation on suction side of main blade and slight separation on suction side of splitter blade near the shroud.	2.5	97.3
6	Increased blade thickness to increase velocities at mid-passage.	Modest separation upstream of main blade and splitter blade, large separation on suction side of splitter blade.	4.0	97.0%	Modest separation on main blade suction surface, possible slight separation on splitter suction surface and slight separation near trailing edge on suction surface of each blade, all near the shroud.	3.0	99.6

* 1 = Weak separation; 3 = Moderate separation; 5 = Strong or large separation.

** This design was selected for fabrication and testing.

\+ Most of the changes are cumulative, i.e., case 4 includes most of the changes for cases 1–3.

Table 5.14 Summary

η *(measured)*	η_{tt} pbCFD™	η_{tt} FINE™/Turbo	ψ *(measured)*	ψ pbCFD™	ψ FINE™/Turbo
0.966	0.9655	98.2	0.79	0.70	0.67

To further model the liquid hydrogen cryogenic pump, the pbCFD™ default grid size (21×71×21) and stretch factor (1.2) was used to continue this study. The inlet conditions were created in CCAD™ [4], with the total pressure and temperature set to the standard reference conditions. The solution took about 20 CPU minutes on a Pentium 400 MHz platform to converge to within 1 per cent of the air-equivalent compressor flow rate. The predicted rotor efficiency was 0.9655, within half a point of the measured data. The rotor head rise coefficient was measured at 0.7; the pbCFD™ value was $\psi = 0.79$. In the computed pbCFD™ solution, there existed recirculation regions near the splitter blade suction surface and in front of the LE, which were observed from previous pbCFD™ computations.

Figures 5.18 to 5.22 show the computed streamlines. Computed streamlines are shown on a number of key surfaces in this pump. Figure 5.18 is located close to the main blade pressure surface and nearly collateral flow is observed, although a small bubble appears to exist just before the leading edge of the main blade. The leading edge is not shown distinctly in this figure, but it is perpendicular to the axis of the plot and is located just aft of the little bubbles shown along the shroud streamline. Similar results are shown near the suction side of the splitter blade, and again, some evidence of a small recirculation zone in front of the leading edge may be observed with larger recirculation downstream of this point, but upstream of the splitter blade (location not shown, but approximately after the separation vortex). Streamline results adjacent to the pressure side are as displayed in Fig. 5.20. The separation zone is small and the distortion in the streamlines is only moderate. Near the mid-channel region, between the splitter and the blade, the flow field is more complex as shown in Fig. 5.21. On the suction side of the main blade, the flow becomes very complex with considerable secondary flow in some regions with local recirculation displayed as revealed in Fig. 5.22. Within the parameters available to influence this design, the condition could not be improved further.

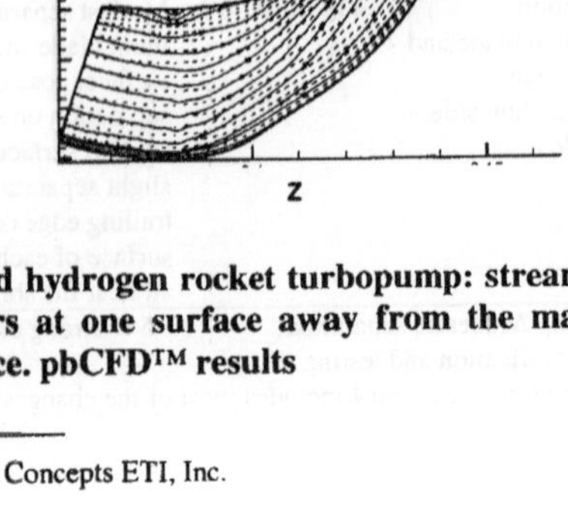

Fig. 5.18 Liquid hydrogen rocket turbopump: streamlines and velocity vectors at one surface away from the main blade pressure surface. pbCFD™ results

4 CCAD™ is a trademark of Concepts ETI, Inc.

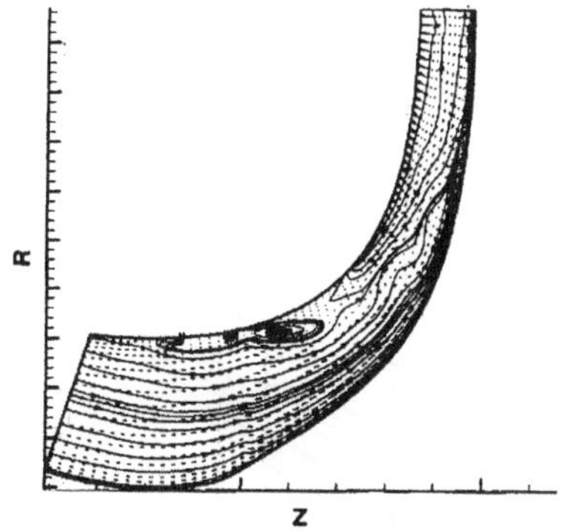

Fig. 5.19 Liquid hydrogen rocket turbopump: streamlines and velocity vectors at one computational surface away from the splitter suction surface. pbCFD™ results

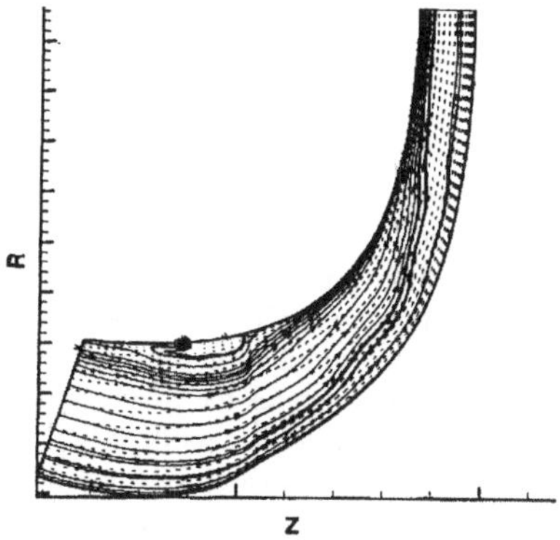

Fig. 5.20 Liquid hydrogen rocket turbopump: streamlines and velocity vectors at one computational surface away from the splitter pressure surface. pbCFD™ results.

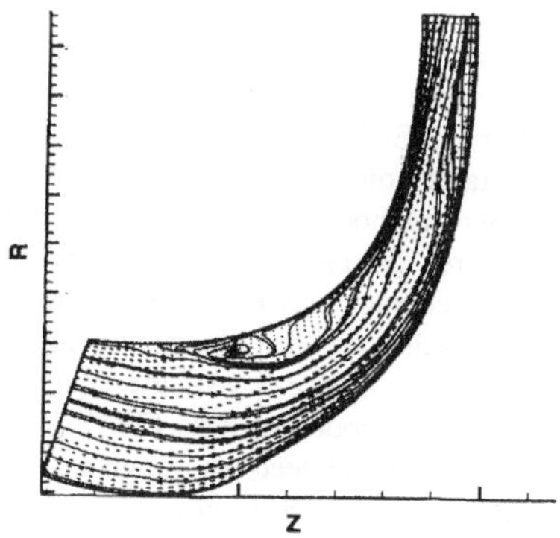

Fig. 5.21 Liquid hydrogen rocket turbopump: streamlines and velocity vectors at mid-channel. pbCFD™ results.

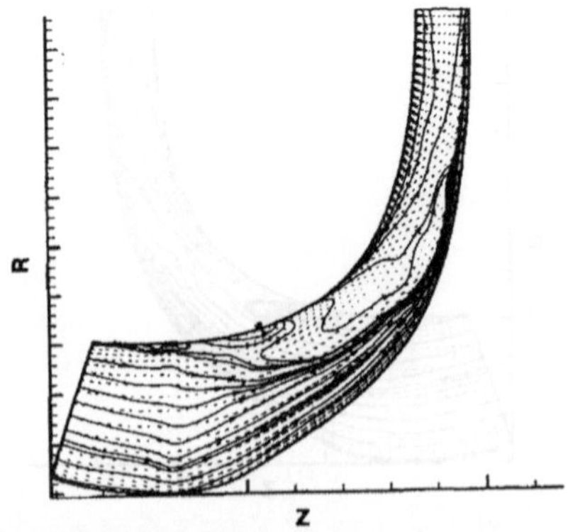

Fig. 5.22 Liquid hydrogen rocket turbopump: streamlines and velocity vectors at one surface away from the main blade suction surface. pbCFD results.

5.6 Summary

A variety of CFD evaluations has been reported covering four of the seven study cases. Data for the other three cases is now added for completeness. While this study is still in its infancy, some comparisons can be made which provide initial understanding of pbCFD™ and the balance between coarse grid and fine grid CFD calculations. Figure 5.23 compares the computed CFD efficiency versus the measured CFD efficiency for pbCFD™ (12 points) and FINE™/Turbo (10 points). Figure 5.24 compares the pressure rise experience. There are still some questions about the best way to calculate both efficiency and pressure rise from measured data and these are being worked separately (methods of averaging) and will be reported further at a later date. We would estimate that the data uncertainty is about ± 1 point, perhaps a bit more. More data and CFD results are needed in Figs 5.23 and 5.24 and such studies are ongoing.

Although the results of Figs 5.23 and 5.24 are broadly acceptable and useful in basic engineering studies, they can still be criticized. It would be nice to have the computational uncertainty reduced to approximately ± one point, or essentially to the level of experimental error. This has not yet been achieved and leaves room for further improvement. Likewise, the variation from either pbCFD™ or FINE™/Turbo in Fig. 5.24 concerning pressure rise or head is not yet acceptable. The uncertainty on this chart is somewhat larger and studies are still being conducted as to find the best method of reporting an appropriate impeller exit average pressure. Nonetheless, these figures do raise a much more basic question: given that a reasonable and nominal level of accuracy has been demonstrated by the coarse grid CFD results, why are not the fine grid results considerably more precise than shown? The question is not about the degree of accuracy or inaccuracy of the coarse grid pbCFD™ results, but rather why respected high-performance codes such as FINE™/Turbo are not showing *more* precision in the computed results. This topic needs thorough examination in the future and the authors of this presentation are conducting such studies.

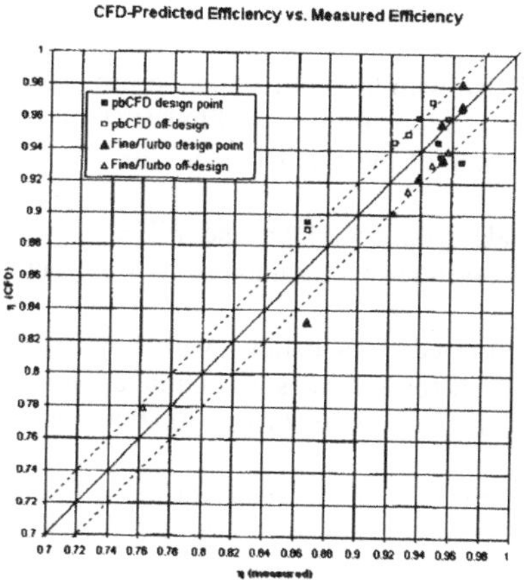

Fig. 5.23 CFD-predicted efficiency versus measured efficiency

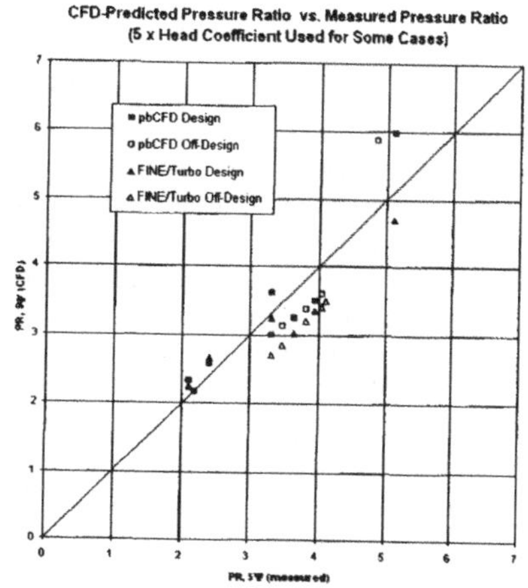

Fig. 5.24 CFD-predicted pressure ratio versus measured pressure ratio (5 x head coefficient used for some cases)

The entire debate about establishing numerical accuracy of CFD codes must be revisited. There is enormous emphasis in the profession on establishing grid independence (a very worthy goal) and hence, many groups concentrate on ever reduced grid sizes to establish this (potentially desirable) objective. If the present study is carefully reviewed, it will be seen that this issue is much more complicated. The results presented here for the pbCFD™ method are not strictly grid independent, although most of the variance due to grid changing has been reduced, eliminated, or controlled. Still, further numerical resolution could be conducted to reduce this type of error. It will require, however, that a fixed condition for y+ near the wall is implemented and that this value is maintained rigorously throughout the computational process. A precise algorithm for doing this has not yet been established in the CFD fraternity. Hence, all methods using a wall log function (pbCFD™ and frequently TASCflow and Fluent products) may be nearly grid independent, but not strictly so at the present time when the wall law is used. Based on the comparative results between a fine grid calculation and the coarse grid calculations in Figs 5.23 and 5.24, it appears that a more important question is what are the relative levels of significance between turbulence modeling, wall shear representation, grid independence, numerical stability/convergence and damping issues within any CFD code.

5.7 Conclusions and recommendations

On the basis of the work presented here, a set of preliminary conclusions has been established. They are as follows:

1. The comparison of computed versus measured efficiencies based on pbCFD™ is good. Nearly all of the results agree within plus or minus two points of efficiency with respect to the measured value. There is a good degree of consistency in the pbCFD™ calculations through all cases.

2. Calculations of rotor total-to-total pressure ratio (or head coefficient for pumps) showed more scatter; further study is needed.
3. Flow patterns computed with pbCFD™ and also with the FINE™/Turbo program were broadly similar between the two codes, with minor differences. FINE™/Turbo gives more exotic detail close to the wall, but the bulk of the characteristics are broadly the same, at least so far as tested at the present time.
4. Several comparative design cases were made with pbCFD™ and FINE™/Turbo to determine what type of design would result; essentially the same results were obtained with each code.
5. Professor John Denton's hypothesis is confirmed. Calculations with approximately 32 000 node points (using pbCFD™) work well for early design optimization studies. Calculations using a code specifically intended for fine grids (about 300,000 nodes) did not give more accurate results.
6. Based on the successful results of pbCFD™, it is evident that the logarithmic law of the wall, as postulated by Prof. Dawes and others, competes very well with other approaches, such as using a low Reynolds number turbulence model equation.
7. Conducting CFD calculations using the logarithmic wall function is quite successful if y+ is set somewhere in the range of 70 to 100 for initial calculations. However, it would be desirable to develop a standardized method that can hold a particular value throughout a flow field calculation, or a specific distribution throughout said calculations, in order to lead towards grid independence in the calculation methods. More development work in this area would be appropriate.
8. An organization using the Dawes BTOB3D code remains on a solid basis but updating to fix known errors is certainly needed. There is no evidence at the present time leading to a substitution for this very fine CFD code.
9. This chapter establishes a comparison between coarse grid and fine grid computational methodologies. Reasonable accuracy has been shown with each, with coarse grid results being of good, basic, engineering accuracy. The question is not whether these calculations will ultimately lead to truly high order precision (they have been demonstrated for basic design iteration studies) but rather why are the fine grid studies not showing greater precision at the present time?

5.8 References

(1) **Denton, J. D.** (1994) *Turbomachinery Design using CFD*, AGARD Lecture Series 195.

(2) **Dawes, W. N.** (1988) *A Computer Program for the Analysis of Three-Dimensional Viscous Compressible Flow in Turbomachinery Blade Rows*, Whittle Laboratory, Cambridge, UK.

(3) **Dawes, W. N.** (1991) The development of a solution adaptive three-dimensional Navier–Stokes solver for turbomachinery, *AIAA* 91-2469.

(4) **Japikse, D., Hinch, D.,** and **Yoshinaka, T.** (1996) The Performance of Low-Solidity Airfoil Diffusers with Centrifugal Compressor Stages at $N_S = 55$ and $N_S = 85$, Advanced Diffuser Consortium, Phase IV Final Report, Concepts ETI, Inc. TM 399, March.

(5) **Hinch, D., Japikse, D.,** and **Yoshinaka, T.** (1997) Range and Performance Enhancement through Diffuser Optimization for the $N_S = 85$ Centrifugal Compressor Stage, Phase V Final Report, Advanced Diffuser Consortium, Concepts ETI, Inc. TM-565, November.

(6) **Japikse, D.** and **Baines, N.C.** (1997) *Introduction to Turbomachinery,* Concepts ETI, Inc., Wilder, VT, and Oxford University Press, Oxford, England. See Section 8.2.

(7) **Japikse, D., Karon, D.,** and **Yoshinaka, T.** (1996) *Evaluation of Rotating Stall at Various Re Levels in a 1½ Stage Process Compressor Rig with Several Vaneless Diffusers*, Final Report, Stability Consortium, Concepts ETI, Inc. TM 400, April.

6

Numerical Determination of Windage Losses on High-speed Rotating Discs

S Romero-Hernández, M R E Etemad, and **K R Pullen**

Synopsis

Determination of the aerodynamic drag, (windage) of rotating enclosed discs is important when designing turbomachinery, motors, generators, and any other high-speed rotating machines. There is limited information, both computational and experimental, for high-speed rotating conditions (30 000 to 60 000 r/min) operating in air. This contribution presents the computational study in this field performed in an effort to fill this gap. The overall results are compared with experiments performed on a purpose-built rig and compare favourably in terms of torque, pressure, and temperature measurements.

The simulation was performed by using a commercially available CFD code (STAR CD®) to generate the grid, verify grid independence, and the general set up of a reliable model that can predict the windage loss on high-speed rotating discs. Using a second order differencing scheme (MARS) and a standard k–ε turbulence model to obtain a solution.

The balance achieved between experimental and computational work in order to obtain an engineering solution is a good example of how a commercial CFD code can become a common practical tool, accessible to all engineers.

6.1 Notation

a	Disc radius	m
b	Radial gap	m
C_m	Torque coefficient for both sides of the disc	
h	Thickness of rotor	m
I_{rot}	Moment of inertia	kg/m^2
K_o	Fraction of disc speed at which the core rotates	m/s
$\dot{m}$	Net mass flow rate through the rig	kg/s
P_w	Power loss	W

R	Radius	m
Re_ω	Rotational Reynolds number, $Re_\omega = \omega\ a^2/\nu$	
s	Axial gap	m
T	Torque	N m
α	Angular acceleration	rad/s^2
μ	Viscosity	kg/m s
ν	Kinematic viscosity, $\nu = \mu/\rho$	m^2/s
ρ	Density	kg/m^3
τ	Shear force on the rotating wall	N/m^2
ω	Angular velocity	rad/s

6.2 Introduction

There is a growing requirement to evaluate the windage losses associated with discs and drums rotating at high speed. Several engineering components such as turbochargers, gas turbines or high-speed electrical machinery would benefit from an improved understanding of the windage losses through increased efficiency and reduced heating. The origin of this heating can be attributed to the drag forces on the rotating component due to viscous friction. In some applications like high-speed electrical machinery, high levels of superposed flow are necessary for cooling purposes. Hence, the determination of the effect of this flow on the windage losses becomes of critical interest.

While the geometries of the inter-disc spacing in an engine are often complicated, the idealized case of flow between plane discs with cylindrical casing forms a useful base for their study. This case is often referred as the enclosed disc.

The pioneering work in this field was done by Von Karman (1921) who used momentum integral equations for turbulent flow and for discs rotating in an infinite environment (the free disc). Substantial research has been performed in the field, mostly concerning low-speed and large-diameter discs; e.g. Daily *et al.* (1) and (2) carried out experiments using liquids and air. It has emerged from this research that there are four basic regimes of flow (laminar and turbulent with merged and separated boundary layers) for a totally enclosed rotating disc. This has prompted the determination of empirical equations for the torque coefficient C_m as a function of the rotational Reynolds number (Re_ω) and the axial gap ratio (s/a). Owen (3), provided a review of the fluid flow for the free disc, rotor stator systems, and rotating cavities. The experiments were performed on air with a disc of a=381 mm (15 inches) and at speeds of up to 4000 r/min. Etemad *et al.* (4) reported the experimental results performed on discs with radii of 50 and 46 mm and speeds up to 90 000 r/min.

The power loss on a disc can be computed from the familiar expression $P_w = T\omega$ and the torque on the disc (both sides) can be calculated from a solution of the momentum equations given in references (1) and (2), as $T = \frac{1}{2} C_m \rho \omega^2 a^5$. The torque coefficient for turbulent flow and merged boundary layers from reference (1) is $C_m = 0.0622/(s/a)^{0.25}\ Re_\omega^{0.25}$, then by combining the previous expressions the power loss for an enclosed disc can be defined as:

$$P_w = 0.0311\left(\frac{s}{a}\right)^{-0.25} \mu^{0.25} \rho^{0.75} \omega^{2.75} a^{4.5} \tag{6.1}$$

Expression (6.1) applies only when there is no superposed mass flow between the rotor and the stator. Ketola *et al.* (5) performed an analysis of the net mass flow between a partially enclosed rotor and a stator. By considering the momentum of the fluid leaving the rim of the disc after flowing in from the centre of the disc, reference (4) re-expressed the work of Ketola (5) to obtain an expression for the power loss due to the superposed flow:

$$P_{inf\ low} = K_o\ \dot{m}\omega^2 a^2 \tag{6.2}$$

Where K_o is the ratio of the angular velocity of the fluid core (between the rotor and stator boundary layers) to angular velocity of the disc ($u_\omega = \omega\, r$). References (2) and (5) assumed a value $K_o = 0.5$ for the turbulent merged boundary layers regime. The total value of the power loss can be predicted by adding the contributions from expressions (6.1) and (6.2).

In the computational aspect, several attempts have been performed (6), (7). Chew (6) concludes that the two equation k–ε model of turbulence has shown to be effective in rotating disc flows. In the numerical work reported in reference (6), conditions were such that compressible effects would be small. This computational work was compared to the experimental results reported in reference (2) and show a good agreement in terms of velocities and torque coefficients. Wild *et al.* (7) used a QUICK discretization scheme to solve the flow around an enclosed rotating cylindrical rotor of a desalination centrifuge. In spite of investigating only low speeds (up to 4000 r/min) they included the effect of the end section between disc and drum. They used a k–ε model and time-averaged mass and momentum conservation equations. However, most of the consulted research is confined to custom-made codes, which can not be easily approached by users.

6.3 Experimental rig and procedure

A purpose-built experimental rig has been designed and commissioned to validate computed work. In this investigation the geometry of study is a hydraulically smooth disc with a radius of a = 47.5 mm and a thickness of h = 10 mm. The radial gap is set to be b = 5 mm and the axial gap has a magnitude of s = 1 mm. There is a forced inflow of air from the inner radius of the disc (bore) flowing radially outwards to the periphery (rim) where it exits the rig, is collected in a manifold and measured by a mass flow meter. The range of superposed flow is $\dot{m}$ = 2 to 13 g/s. A schematic of the rig is depicted in Fig. 6.1.

The testing methodology has been based on the inertia run-down technique to evaluate the windage losses. This method relies on the determination of the torque, T, due to windage acting on the rotor. After bringing the test rotor unit to the constant desired speed, the rotor unit is then quickly moved away from the turbine drive so that the rotor unit starts to decelerate without any external influence. The only restraining torque during the run-down is the windage losses on the rotor and the small contributions from the windage losses of the coupling and the mechanical losses on the bearings.

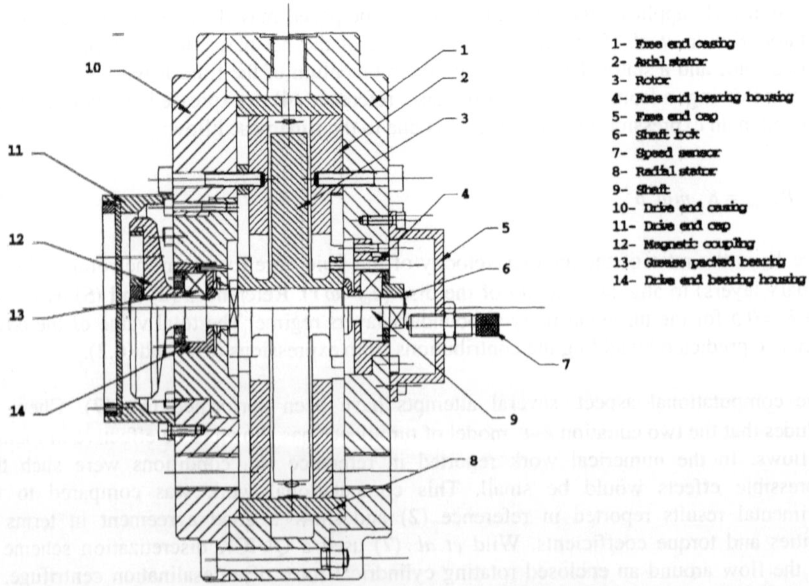

Fig. 6.1 Cross-section of experimental rig

By using Newton's second law for a rotating body ($T = I_{rot}\ \alpha$), the total torque acting on the test rotor is determined, I_{rot} is the moment of inertia of the rotating components and α is the angular deceleration. Therefore, by recording the deceleration after the magnetic coupling is decoupled, the torque exerted on the rotor is computed. The inertia run-down method determines the total torque acting on the rotor, therefore the energy losses on the bearings and coupling must be subtracted from the total through equation (6.3)

$$P_{windage} = P_{total} - P_{coupling} - P_{bearings} \tag{6.3}$$

A full description of the methodology and experimental results can be consulted in reference (8) and PhD work due for submission 1 May, 2001.

6.4 Computational model

For the study of the windage it was decided to model a sector of the gap between rotor and stator. The flow is assumed to be steady, turbulent, and axi-symmetric and therefore, the problem can be treated as two-dimensional in the sense that gradients in the angular direction are neglected, but all three velocity components are taken into account. The turbulent kinetic energy k, and its dissipation rate ε are determined using the standard k–ε model. In this model the high (turbulent) Reynolds number forms of the k and ε equations are used in conjunction with an algebraic wall function representation of flow within the boundary layers.

6.4.1 Boundary conditions

All the fluid cells are given a positive spin equal to the rotational speed of the rotor. The stator walls have a negative spin of the same magnitude, consequently in a static frame of reference the stator walls remain static and the rotor walls are rotating with the cells. In order to account for the angular component of speed a cyclic symmetry condition was applied on the upper and lower faces of the domain. To reduce the computational time, only half of the rotor-stator geometry was modelled, and a symmetry plane at the middle of the drum section was specified. Since the flow enters radially to the system (at the bore), it was necessary to give it also a negative spin so in the static frame of reference the radial condition would be achieved. The value for the inlet velocity, U_{in}, is computed from the mass flow rate in study; the values for the turbulence intensity and its dissipation rate can be obtained by using the expressions:

$$k = \frac{3}{2}(I^2 U_{in}^2) \tag{6.4}$$

$$\varepsilon = C_\mu^{3/4} \frac{k^{3/2}}{l} \tag{6.5}$$

Where I is the turbulence intensity, typically between 1 and 5 per cent, C_μ is a dimensionless constant, usually taken as 0.09 and l is the characteristic length scale taken as 10 per cent of the inlet length (Versteeg, *et al.*, **(9)**. A pressure boundary condition was applied at the exit of the domain, with no user conditions for temperature and turbulence intensity, hence the boundary values for temperature, k and ε are calculated by the code on the assumption of zero flux gradient along the streamlines intersecting the boundary surface.

Figure 6.2 depicts the location of the aforementioned boundary conditions on the structured meshed for the CFD domain.

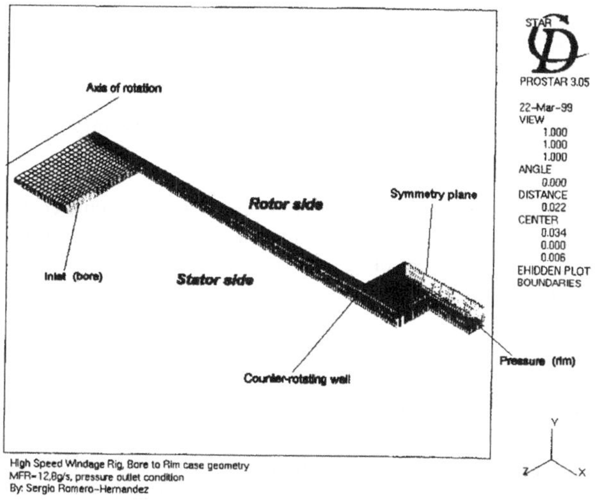

Fig. 6.2 Mesh and boundary conditions

6.4.2 Mesh refinement and numerical solution

The numerical solution was performed using the finite volume program STAR CD®. The SIMPLE (Semi–Implicit Method for Pressure-Linked Equations) algorithm was used. It is an iterative technique, which guesses a value for the pressure and then corrects it during the next iteration. The multidimensional second-order differencing scheme MARS (Monotone Advection and Reconstruction Scheme) was used. The present model was set to solve for velocities, pressure, kinetic energy, viscous dissipation, temperature, density, and viscosity, and therefore compressibility effects are taken into account.

Mesh refinement was conducted in three stages. First, computations were performed for a non-uniform structured mesh composed by 2260 elements, great care was taken to ensure that the aspect ratio from cell to cell never exceeded a value of 1:5 as recommended by Versteeg *et al.* (9). Two successive refinements were performed first in the radial direction to produce a 4520 element mesh, and then in the axial direction resulting in a 9040 element mesh.

All three runs were compared in terms of pressure and temperature on the stator wall. Figure 6.3 shows the non-dimensional pressure and temperature against the non-dimensional radial distance for the three meshes, note the axis scales.

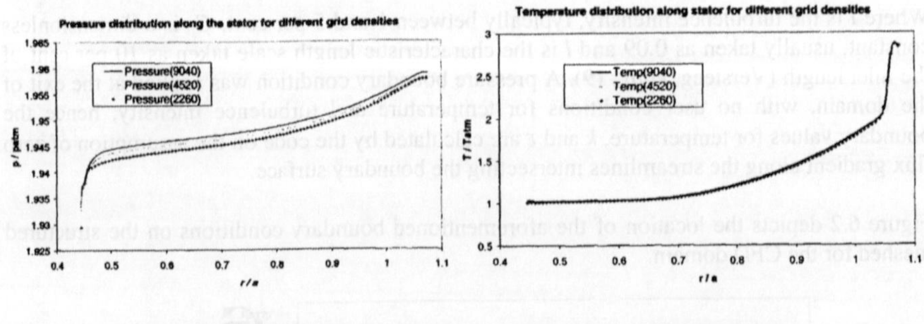

Fig. 6.3 Effect of grid refinement on pressure and temperature

The maximum percentage difference between the initial mesh and the refined one was 0.1 per cent for the pressure and 3 per cent for the temperature. Hence the original mesh of 2260 elements is considered to be mesh independent.

The total torque acting on the rotor wall (windage torque) can be calculated from the CFD runs by taking the circumferential shear force acting on each of the cells on the rotor wall. Then if the radial position of the centre of gravity of each cell is known the torque for each cell can be computed ($T_{cell} = \tau_{cell} r_{cell}$). Finally, the addition of all cell torques will be the total torque on the rotor.

6.5 Results and discussion

The windage losses computed are compared to the experimental data and results calculated using equations (6.1) and (6.2). The values for the windage losses at 60 000 and 30 000 r/min are presented in Fig. 6.4.

It can be seen that the numerical predictions of the windage losses are in good agreement with the experimental results. The computed predictions show the same trend as the experiments with a maximum percentage difference of 7 per cent at 60 000 r/min and 8 per cent at 30 000 r/min. In both cases, the maximum error occurs for the lower values of mass flow rate suggesting that the k–ε model is not suitable for cases where the boundary layer flow is dominated by the drag exerted on the rotor.

The calculated values from equations (6.1) and (6.2) are higher than the experimental results by 8 to 30 per cent for 60 000 r/min and 25 to 67 per cent for 30 000 r/min. The disagreements inferred to come from the fact that these equations were developed from experimental data at low-rotational speeds and large discs. The assumption that the angular velocity of the fluid core would be half the speed of the disc was based on the boundary layer profiles derived for the case of non superposed flow, and show only to be valid for small values of superposed flow. Furthermore, the angular velocity of the core is reduced by the superposed flow.

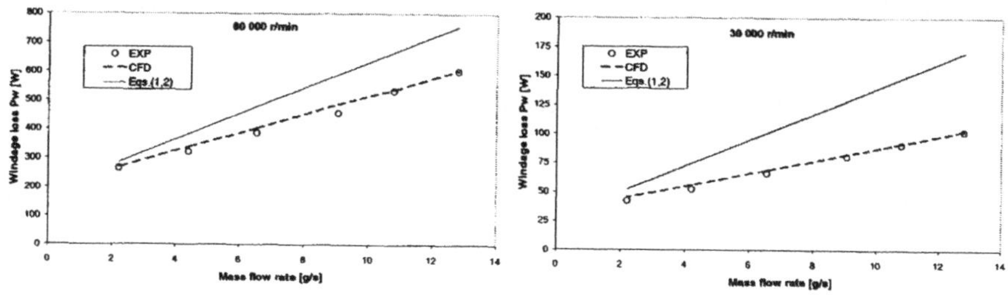

Fig. 6.4 Comparison of computed, measured, and calculated windage losses against mass flow rate

It can also be seen that the extra momentum caused by the superposed flow has a linear effect on the windage losses, just as predicted by Ketola *et al.* (5) on equation (6.2).

6.6 Conclusions

A reliable CFD model for high-speed rotating discs with inflow has been developed using a commercial code. A satisfactory agreement is achieved in terms of windage loss, pressure, and temperature. The solution is based on the standard k–ε model of turbulence and a second order differencing scheme (MARS). Future work will include a through analysis of the effect of mass flow rate on windage losses. It is expected that a correction of the value of K_o will improve the calculated predictions. The agreement achieved between experimental and computational work in order to obtain an engineering solution is a good example of how a commercial CFD code can become a common practical tool, accessible to all engineers.

6.7 Acknowledgements

The authors wish to thank the National Council for Science and Technology (CONACYT) for the financial support to Mr. Romero-Hernández during his studies.

6.8 References

(1) **Daily, J. W.** and **Nece, R. E.** (1960) Chamber dimensions effects on induced flow and frictional resistance of enclosed rotating disks, *ASME J. Bas. Engng*, March.

(2) **Daily, J. W., Erndst, W. D.,** and **Asbedian, V. V.** (1964) Enclosed rotating disks with superposed throughflow, MIT Hydrodynamics Laboratory, Report No. 64, April.

(3) **Owen J. M.** (1984) Fluid flow and heat transfer in rotating disc systems, Proceedings of the *International Centre for Heat and Mass Transfer,* **16**, pp. 81–103.

(4) **Etemad, M. R., Pullen, K. R., Besant, C. B.,** and **Baines, N.** (1992) Evaluation of windage losses for high-speed disc machinery, *Proc. Inst. Mech. Engrs*, **206**.

(5) **Ketola, H. N.** and **McGrew, J. N.** (1968) Pressure frictional resistance, and flow characteristics of the partially wetted rotating disk, *ASME J. Lubric. Technol.,* April.

(6) **Chew, J. W.** and **Vaughan, C. M.** (1988) Numerical predictions for the flow induced by an enclosed rotating disc, *ASME Gas Turbine and Aeroengine Congress*, Amsterdam.

(7) **Wild, P. M., Djilali, N.,** and **Vickers, G. W.** (1996) Experimental and computational assessment of windage losses in rotating machinery, *ASME J. Fluid Engng,* **118**, 116–122, March.

(8) **Romero-Hernández, S.** (1998) *Determination of windage losses on high-speed rotating discs with superposed flow*, MPhil to PhD transfer report, Imperial College, London.

(9) **Versteeg, H. K.** and **Malalakasera, W.** (1995) *An Introduction to Computational Fluid Dynamics*, Addison Wesley Longman.

7

Computational Fluid Dynamics for Gas Turbine Combustion Systems – where are we now and where are we going?

K Menzies

Synopsis

Computational Fluid Dynamics (CFD) is being used with increasing regularity in the design and development of combustion systems for gas turbine engines. It offers the potential for more rapid development of the combustion system with fewer experimental tests and lower development costs. However, in order to realize this potential we must be able to reproduce the physical processes with a level of accuracy sufficient to predict the parameters of interest to the combustion engineer. In this chapter, we consider the current capabilities of CFD for gas turbine combustion problems from the perspective of what a typical user working on an engine project needs to be able to predict. Out of the various areas influencing solution accuracy and reliability we concentrate on some issues in physical modelling and numerical accuracy. In particular, we consider the minimum standard of turbulence and combustion modelling required to achieve meaningful results and recommend numerical methods to deliver a level of numerical accuracy that repays the use of sophisticated physical models. We also look at current research in these areas to examine the potential improvements that can be achieved to extend the use of CFD as a part of the combustion engineer's toolkit.

7.1 Introduction

The modern gas turbine engine is a highly complex feat of engineering as illustrated by the cut-away of a typical large civil turbofan in Fig. 7.1. In the increasingly competitive global aerospace and power generation markets, it is vital to be able to design and develop engines to ever shorter timescales and lower cost, while meeting more stringent operating requirements. In this environment CFD now plays an increasingly important role in allowing the engineer to simulate any gas flows within the engine.

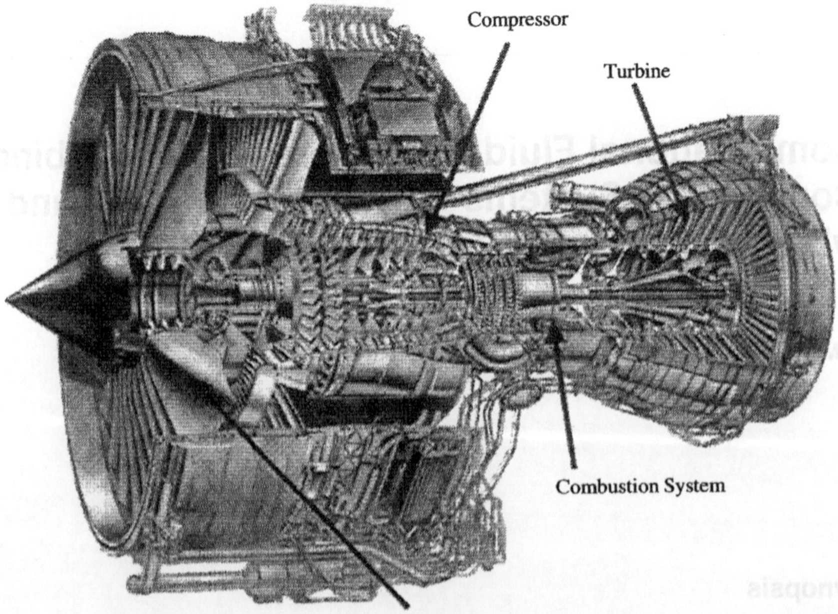

Fig. 7.1 Typical modern aero gas turbine

In this chapter we shall survey the use of CFD in the design of combustion systems. We clearly cannot cover all issues in the use of CFD and so are forced to omit some material, in particular discussion on liquid fuel modelling and boundary conditions.

7.2 The gas turbine combustion system

A typical combustion system for an aero gas turbine is shown in Fig. 7.2. Air flows from the compressor through the diffuser before entering the combustion chamber. Fuel is added through the fuel injector; the fuel-air mixture reacts inside the combustion chamber. The reacted mixture exits through the turbine downstream of the combustion system. The combustion chamber walls contain holes to control the air admission into the chamber, to both modify the temperature distribution within the chamber and provide cooling for the combustor walls. The aero combustor shown is an example of a diffusion flame combustion system where the fuel and air enter the chamber in distinct streams and combustion takes place in the mixing layers. An alternative technology is to use premixed combustion systems where the fuel and air is mixed as uniformly as possible before entry to the combustor; this technology is extensively used for combustion systems for stationary power generation and offers the potential for lower levels of pollutant emissions.

The combustion system must be designed to meet a number of goals, some of which are conflicting. We will typically want to meet targets on at least the following criteria:

- Combustion stability, to ensure that the flame is not extinguished during transient or low-power operation;

Relight capability, ensuring that the flame can be re-ignited if it is extinguished to a guaranteed altitude; this places limits on the minimum size of the primary zone of the combustor;

Emissions performance, to reduce the production of combustion-generated pollutants such as oxides of nitrogen, carbon monoxide, carbon dioxide, smoke, or unburned fuel. These may have detrimental effects on the environment and some are the subject of international legislation. Generally, we wish to have a smaller combustion chamber to minimize production of oxides of nitrogen;

- Reduce combustor length to directly minimize mass and to reduce the length of the engine shafts;
- Guaranteed combustor life before mechanical damage accumulates;
- Exit gas temperature distribution which enters the turbine;
- Overall pressure losses.

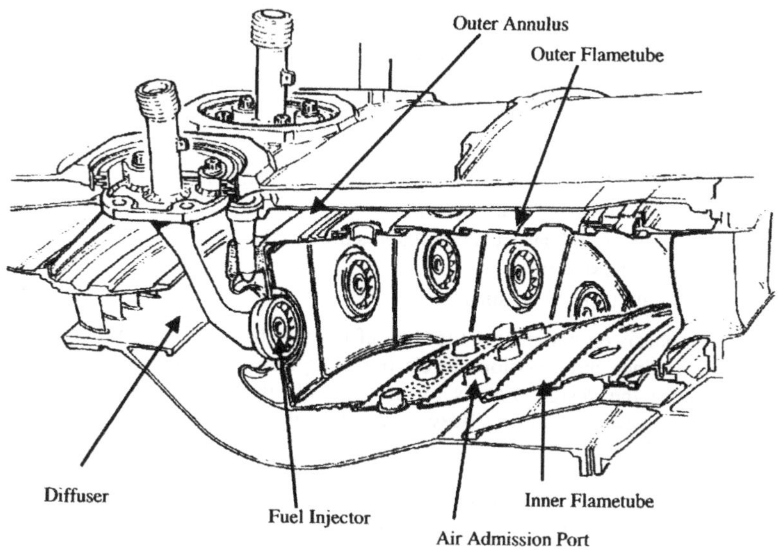

Fig. 7.2 Gas turbine combustion system

At its most arduous operating point the combustion system may be functioning with an inlet air stream at a pressure of around 30 atmospheres and temperature of 800 K; peak flame temperatures may reach 2600 K. Given these conditions we may see that taking experimental measurements is extremely difficult. Combustor test rigs generally can operate only at compromised conditions, giving limited data at lower pressures (perhaps 10 atmospheres) or more detailed data at 1–5 atmospheres. In combustor development by experimental methods, we would use high-pressure rig testing (10 atm) which allows us to measure the exhaust gas temperatures and composition, along with measurements of the combustor wall temperatures. Full conditions can, of course, be achieved by running engine tests but even less information is available here, generally only metal temperatures, and this requires the existence of a suitable engine which will not be the case in the early stages of the design. Both rig and engine testing are also expensive and slow; if we wish to reduce development timescales and costs we must employ alternative tools to aid our design process.

7.3 The role of CFD in the design of the combustion system

Traditionally, CFD has been used less extensively in combustion system design than in some other areas of the engine, due to the complex nature of the flow. The combination of highly three-dimensional flow, large-scale turbulence, and highly exothermic chemical reactions has provided a severe test of our physical understanding in constructing tractable approximations as well as placing great demands on computational resources to solve the resulting numerical problem. However, advances in turbulence modelling, combustion modelling, and numerical methods, coupled with the growth in computer power available in relatively modest desktop machines, now makes CFD analysis of combustion systems viable as a routine part of the design process. Despite the limiting assumptions built in to some of the physical models used in the past, it has been possible to perform useful calculations for combustion problems as illustrated by references (**3**), (**8**), and (**22**), but we are now in a position to significantly reduce timescales and costs by the use of CFD.

The objective for the combustion engineer of using CFD must be to reduce design and development timescales and costs. We can achieve this if we can predict the major parameters of interest to the engineer reliably, consistently, and on time, thereby reducing the level of rig and engine testing required to prove the design. We may wish to predict, among other features,

- The exit temperature distribution from the combustion chamber;
- The pressure losses incurred in the system;
- The levels of gaseous and particulate pollutants produced;
- Thermal loads on the structure;
- Stability and ignition characteristics.

With increased emphasis on component life, fuel efficiency (both affecting the total cost of ownership of the engine) and emissions levels (resulting from international legislation) we can clearly influence the design process by use of CFD if the methods are capable of predicting parameters such as wall heat fluxes and pollutant formation. However, as we shall see, this is not a trivial exercise!

We must also note that these CFD methods should be suitable for use by suitably-trained engineers, not just 'CFD specialists'. We must ensure however, that the CFD-using population is sufficiently well trained to assess their results in the light of the limitations of the physical models and numerics employed.

7.4 Formulation of the problem

The conventional approach employed for CFD analysis of combustion systems is to write the governing Navier–Stokes equations and chemical species transport equations in density-weighted (Favre) averaged form. We assume that the flow is at a sufficiently low Mach number that the only density variations arise through combustion. We further assume that the flow is steady, giving the following governing equations in Cartesian tensor form, where ϕ'' is the Favre fluctuation and $\tilde{\phi}$ represents a density weighted mean value:

Continuity:

$$\frac{\partial}{\partial x_i}(\bar{\rho}\tilde{u}_i) = 0$$

Momentum:

$$\frac{\partial}{\partial x_i}\left(\bar{\rho}\tilde{u}_i\tilde{u}_j\right) - \frac{\partial}{\partial x_i}\left(-\overline{\rho u_i'' u_j''}\right) = -\frac{\partial p}{\partial x_j}$$

Species transport:

$$\frac{\partial}{\partial x_i}\left(\bar{\rho}\tilde{u}_i\tilde{Y}_j\right) - \frac{\partial}{\partial x_i}\left(-\overline{\rho u_i'' Y_j''}\right) = \bar{\dot{\omega}}_j$$

We then obtain a number of unclosed terms: the Reynolds stresses $-\overline{\rho u_i'' u_j''}$, the scalar fluxes $-\overline{\rho u_i'' u_j''}$, the mean chemical source term $\bar{\dot{\omega}}_j$, and the mean density $\bar{\rho}$, as discussed in (**12**). Each of these must be modelled and the assumptions employed in this modelling control which of the aspects of combustor performance we can predict.

7.5 Turbulence modelling

The majority of combustor calculations reported in the literature employ the high Reynolds number k–ε turbulence model (**13**), written in density weighted form; we assume that the effects of the density fluctuations are accounted for in the averaging process. This allows us to model the Reynolds stresses in terms of the mean rate of strain and a 'turbulent' viscosity; the scalar fluxes are closed through a gradient transport hypothesis. This model is robust and can produce good results; however, it is unable to predict features such as the modification of mixing rates by strong mean swirl (as is found in the efflux of the fuel injector), or the shift in mean velocity profile in the annuli resulting from stress relaxation. These are all effects resulting from stress anisotropy or redistribution which cannot be captured when the stress field is characterized by the (scalar) turbulence kinetic energy and interacts with the mean flow through the turbulent viscosity. The current alternative method is to employ a second-moment closure that formulates modelled transport equations for the individual stress components. These Reynolds stress models have been used to good effect on some complex flows (**20**) and are able to capture a wider range of physical phenomena.

Despite these concerns, the k–ε model can produce results that agree well with experimental data even for flows as complex as those in combustion chambers, as illustrated in Figs 7.3 and 7.4 for a research combustor; the experimental results are from (**2**). However the predictions for diffuser systems are less good with the k–ε model: while the mean pressure loss may be predicted reasonably correctly the velocity profiles may be severely in error. In particular the standard k–ε model predicts a growth in turbulence energy (and corresponding increase in turbulent viscosity) at the impingement point on the head of the chamber where the flow decelerates. This is then transported around the head of the combustor, corrupting the velocity profiles predicted in the annuli. In contrast a Reynolds stress model will correctly predict no growth in turbulence energy through the deceleration of the flow (instead the energy is redistributed among the stress components), leading to more realistic predictions of the flow around the combustor head.

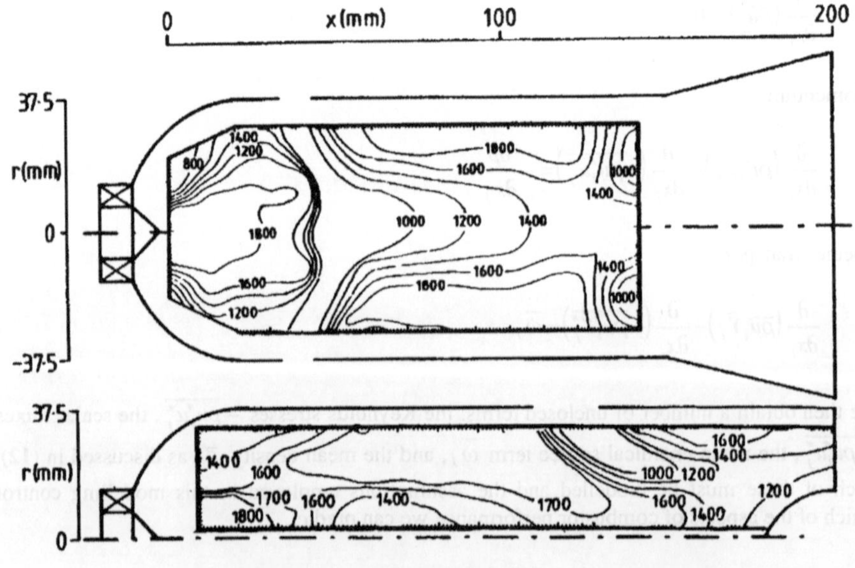

Fig. 7.3 Research combustor temperature measurements

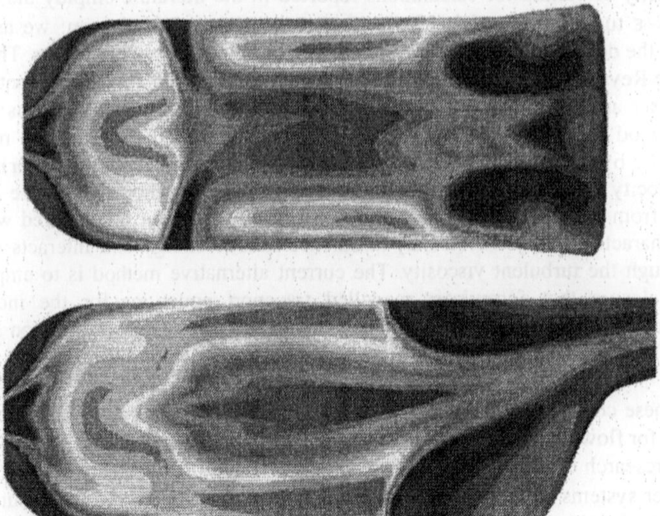

Fig. 7.4 Calculation of research combustor

The use of Reynolds stress models in combusting flows is less well developed as a result of the relatively smaller amount of effort put into developing second-moment closures suitable for reacting flows, and the difficulty of taking measurements for model validation and development. The stress transport equations are again written in density-weighted form but now additional terms appear for combusting flows which require modelling. Of the available models, that described in **(11)** appears to be one of the most reliable and has produced good results for reacting flow. The scalar fluxes must again be modelled and this may be done using the second-moment closure also described in **(11)**. This level of closure allows us to capture effects such as counter-gradient diffusion observed in premixed flames which cannot be represented by lower order closures.

The Reynolds stress and scalar flux closures described represent the current best practice for application to practical gas turbine combustor calculations, since they balance physical realism with computational economy. However, we are still limited by the fundamental assumptions of all Reynolds averaging methods whereby we have characterized the turbulence field by a single length and time scale. This is likely to be a significant compromise particularly in flows where the effects of solid walls are important, such as the modelling of cooling films or diffuser performance. In such cases we may wish to move to a more comprehensive treatment of the turbulence structure that captures at least some of the range of eddy scales and so the energy cascade. The only method currently offering promise in this area for engineering flows at high Reynolds number is the Large Eddy Simulation (LES) approach. Here, we perform an unsteady calculation on a grid sufficiently fine to resolve the large eddy structures and simulate only those that cannot be resolved on the grid (which should be that part of the flow within the turbulence inertial subrange). This method is attractive for a number of reasons for calculating the flows in combustion systems:

- The inherently unsteady nature of the calculation and capture of multiple length and time scales should allow the calculation of structures such as cooling films which are poorly predicted by many Reynolds-averaged techniques. The decay of the cooling film is controlled by the rate of mixing between the film and the hot mainstream; this mixing process is governed by large-scale vortex motions in the interface leading to entrainment of hot gas into the film. These processes should be more amenable to prediction by LES than by Reynolds-averaged methods which should improve the data that can be supplied to combustor life calculations.
- LES allows us to directly simulate the larger turbulence scales which will be most influenced by the geometry and boundary conditions. The smaller, modelled, scales are assumed to be more universal and therefore may be more properly approximated by simple 'universal' models. This should further improve diffuser predictions where small changes in the geometry may lead to significant changes in performance and the flow characteristics are dominated by the structure of the near-wall flow.
- In attempting to reduce emissions from the combustion system, most approaches now favour use of a fuel-lean flame. This introduces issues of flame stability and the possible generation of destructive acoustic waves in the chamber. These can contain sufficient energy to damage the combustion system. These instabilities take the form of large-scale unsteady motions in the combustion chamber which should be better captured by LES thus allowing investigation of potential acoustic problems at the design stage instead of using expensive rig testing.

However, we should not underestimate the challenges involved in making LES a design tool. Relatively little work has been performed on LES for combustion, some exceptions being for example **(1)**, **(5)**, and **(7)** and so issues remain over the precise treatment of combustion in an LES framework. While we assume that the unresolved subgrid turbulence scales are represented by relatively simple universal models, there has been little comparative testing of different subgrid scale models in complex engineering flows. We also need the calculations to be rapid enough to provide timely information in a design environment which will place demands on numerical methods and computer hardware.

7.6 Combustion modelling

We must also determine the values of the other unclosed terms, the mean density and mean chemical source term, in order to model our combusting flow. The approximations that we employ here significantly affect the range of operating conditions that we may calculate and the parameters that we may confidently use from the prediction.

We must note that there are different regimes of turbulent flames; these have been characterised by Borghi (4) into the types shown in Fig. 7.5. The discriminator is the Damköhler number: the ratio of turbulent to chemical timescales. At high Damköhler numbers the chemical reactions are fast compared to turbulent mixing times; at low Damköhler numbers the chemical reaction times for the heat release reactions are comparable to the turbulent mixing times.

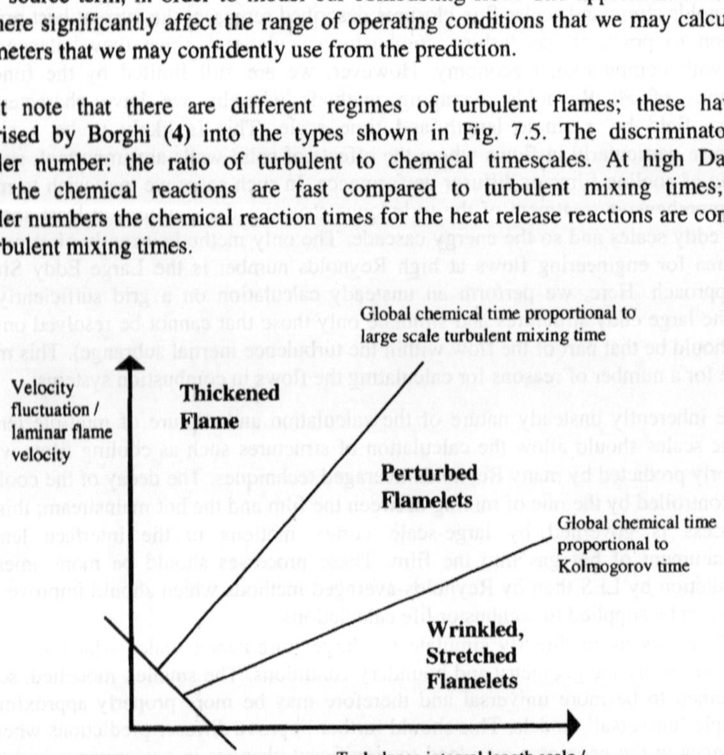

Fig. 7.5 The regimes of turbulent diffusion flames (Borghi diagram)

7.6.1 Modelling combustion chemistry

Practical gas turbines employ complex hydrocarbon fuels, typically aviation kerosene, diesel, or natural gas. While the chemical reaction mechanisms governing the combustion of natural gas are now established there is less certainty about the combustion of kerosene. Moreover, the detailed chemical reaction mechanisms that do exist are highly complex: a typical model kerosene mechanism employs 92 chemical species and 460 reactions (18) giving a large system of coupled stiff ordinary differential equations. This level of complexity is not tractable for use within three-dimensional turbulent flow calculations.

For diffusion flames if we confine our attention to the high Damköhler number regime and further assume that all species diffusivities are equal then we may characterize the combustion process by the value of a single conserved scalar which eliminates the mean chemical source term (**12**); this conserved scalar is often taken as the mixture fraction, ξ. We are then left with the calculation of the mean density from the mixture fraction. A similar approach can be used for premixed flames where the conserved scalar is a measure of reaction progress.

The two most commonly used methods for characterizing the thermochemistry in high Damköhler number diffusion flame calculations are to either assume chemical equilibrium or to employ a laminar flamelet description in a pre-processing stage, taking the results into the CFD code. The former method has been used extensively (e.g. (**8**), (**22**)) and is attractive because it does not require knowledge of the chemical reaction mechanism. However, although it is able to give reasonable predictions of density and temperature this method is not suitable for predicting many combustor emissions since the measured values are far from the equilibrium levels. An improvement is to employ the so-called laminar flamelet model. This approach requires us to calculate the properties of a laminar flame at the conditions needed by the combustor calculation which is only possible when we have a chemical reaction mechanism available. The laminar flamelet approach does permit better predictions of species such as CO than equilibrium, but is still limited by the decoupling of the chemistry and the flow. It is, however, the method of choice for practical calculations at present with the understanding that it is only applicable for high-power predictions. In the case of premixed flame modelling the usual approaches are to employ either a premixed laminar flamelet approach, a global one-step chemical reaction mechanism (which will be very crude) or a global two-step mechanism.

While using a detailed chemical reaction mechanism is prohibitively expensive within the CFD calculation, a promising approach is to employ a reduced chemical mechanism. This describes the combustion process in terms of perhaps four key reactions involving five scalar variables, which allow prediction of the major species and does not invoke the assumption of high Damköhler numbers. This method has been demonstrated to be effective for laminar flames (**14**) and is now being adopted for turbulent flame calculations although this application is complicated by the turbulence-chemistry interaction treatment (see the next section). This is the 'next-step' method for modelling the combustion process, although it requires specification of a reduced mechanism. There is some potential for these to be calculated automatically via a method such as the Intrinsic Low-Dimensional Manifold (**19**), but given the level of reduction required this is a daunting task. An attractive alternative is the In-Situ Adaptive Tabulation method of Pope (**21**), which is claimed to offer the ability of treating more complex chemistry in turbulent flow calculations, although this is still to be demonstrated for practical engineering calculations. Some form of chemical reaction scheme coupled with the flow solution is essential for predicting low-power operation, particularly emissions; at present the reduced mechanism approach appears most promising. This will also permit a unified methodology for diffusion and premixed flames.

Prediction of soot emissions is especially problematic in the combustor. The kinetics of soot formation and oxidation is not yet well understood for complex hydrocarbon fuels. However, some progress has been made in understanding the kinetics (**18**); there has also been application of a model based on a laminar flamelet description of combustion in (**6**). The results are very encouraging but more development is needed before it becomes a routine tool; this is an important research need.

7.6.2 Modelling turbulence-chemistry interaction

In a combusting flow, variables such as density and temperature are highly nonlinear functions of the chemical state. We cannot simply calculate the mean value of the thermochemical variables from the mean values of the solved scalars; we must account for the distribution of the scalar fluctuations. This is accomplished by use of the probability density function (pdf) of the scalars. In the case of a single conserved scalar we simply have

$$\tilde{\phi} = \int_0^1 \phi(\xi)\tilde{P}(\xi)\mathrm{d}\xi$$

where $\tilde{P}(\xi)$ is the density-weighted pdf. A number of forms for $\tilde{P}(\xi)$ have been used in the literature; a common choice is to take $\tilde{P}(\xi)$ to be a beta function defined from the mean mixture fraction and the mixture fraction variance **(8)**, **(22)**. When using a laminar flamelet combustion model the values of $\phi(\xi)$ are obtained from the flame calculation.

If we wish to employ chemistry more complex than simply a single parameter tabulation, then we have the problem of defining a multi-dimensional pdf with integration limits that depend on the species concentrations. One approach to multi-step chemistry in turbulent flows is the eddy breakup model **(24)** which avoids this problem completely, allowing the use of single-step or two-step chemistry. It simulates the effect of turbulence on the combustion process by merely using the mean mixture temperature to calculate reaction rates which are then modified by a turbulent mixing controlled step. This model is essentially intuitive and contains a model constant that may be tuned to modify the results. It is still restricted to the mixing-controlled (high Damköhler number) regime. A more rigorous approach which is applicable beyond the high Damköhler number limit is to derive a transport equation for the pdf and solve this as part of the flow calculation; this circumvents our problem of how to specify the pdf *a priori*. This pdf transport equation contains terms which require modelling; in some cases there are no ideal models. However, the chemical reaction mechanism appears without approximation in this formalism which makes it ideally suited for the reduced reaction schemes discussed above. Since turbulence and reaction are coupled here the method is able to treat all operating conditions, limited only by the fidelity of the reaction mechanism. This method is being actively developed and is undoubtedly set to become the method of choice for reacting flow calculations allowing a wider range of calculations to be performed.

We must emphasise that in flows such as are found in combustion chambers, with high levels of turbulence, the interaction between turbulence and reaction must be accounted for. Simply using the mean flow properties to calculate the thermochemical data without modification can lead to large errors **(10)**. We should beware of codes that offer arbitrarily complex chemistry without coupling reaction to turbulence!

7.7 Numerical methods

The accuracy and reliability of our CFD solution is also influenced by the numerical methods employed. We can only repay the investment in complex turbulence and combustion models if the numerics are sophisticated enough to discriminate between models; we cannot afford to lose the beneficial effects of improved physics through poor numerics.

Historically, many practical engineering CFD calculations have used methods such as hybrid differencing or similar discretizations **(8)**, **(22)**. These methods have the advantage of stability but at the expense of unacceptably high levels of numerical diffusion on practical grids when the flow is not aligned with the grid. The development of schemes such as QUICK **(16)** allowed solutions with much lower levels of artificial diffusion but simultaneously with no guarantee of boundedness; some examples of applications to combustor flows are given in **(20)**. The development of bounded second-order schemes such as **(9)**, **(26)** for low Mach number flow calculations has removed the need for any compromise between diffusion and boundedness; these methods can also be implemented in a stable manner as illustrated by **(26)**. We can only repeat the sentiments of Leonard and Drummond's 'manifesto' on differencing schemes **(17)** that there are far better alternatives to first-order schemes and all practical calculations should be using these alternatives. In application to combustor flows we are now able to employ accurate discretization on all variables including those that are physically positive-definite such as the turbulence energy or mixture fraction. The calculations shown in Fig. 7.4 employ the van Leer **(25)** scheme implemented in the manner suggested by **(11)** and **(26)** for all variables and this method is in routine use on practical calculations. For methods such as LES we may need to investigate more sophisticated discretization schemes.

Conventionally, the CFD user sets up the computational grid for their problem before performing the flow calculation. In order to reduce user workload and improve numerical accuracy we have pursued the concept of solution-adaptive grids to allow the grid to redistribute to resolve the underlying solution **(15)**. This permits higher numerical accuracy than can be achieved by manual definition of a similarly-sized grid and also delivers a faster solution turn-around than a finer grid that may deliver similar accuracy. This faster turn-around allows CFD to become more responsive in a design environment. Solution-adaptive grid redistribution had been applied to calculations of practical combustor geometries, such as **(15)** and demonstrated their applicability, as shown in Fig. 7.6, using an adapted grid of the same size as the results in Fig. 7.4.

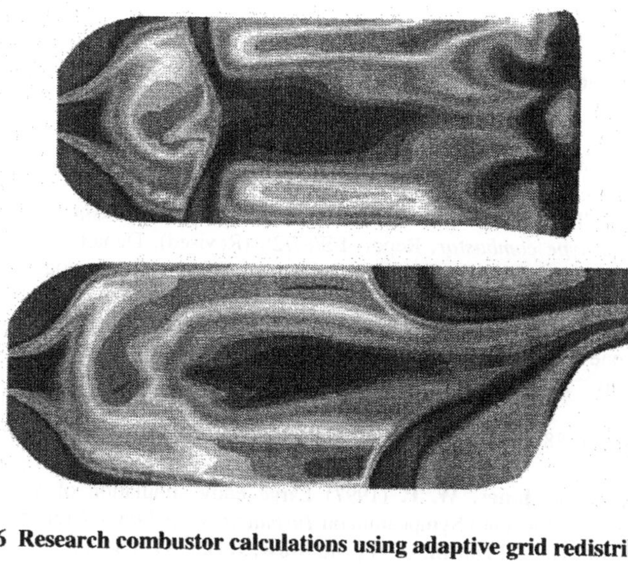

Fig. 7.6 Research combustor calculations using adaptive grid redistribution

For the future we must routinely employ adaptive techniques for practical calculations, perhaps combining adaptive redistribution and refinement, with the underlying flow solution calculated using non-diffusive and bounded numerical methods. This will allow the best utilisation of finite computer resources to produce numerically accurate solutions in timescales relevant to the design process.

7.8 Conclusions

In the preceding sections we have attempted to highlight some of the issues involved in employing CFD in a practical design environment for gas turbine combustion systems. We have seen how current methodologies based on either k–ε or second-moment turbulence models and laminar flamelet descriptions of combustion chemistry are able to deliver realistic results for high-power operation. The next steps in combustor prediction are to employ the transported pdf method with (simplified) multi-step chemical kinetics to allow prediction of low power operation and kinetically-controlled emissions. We must also extend our turbulence modelling methodology to encompass LES to allow better prediction of complex wall-bounded flows and unsteady phenomena; the latter are particularly difficult to investigate experimentally. Eventually the aim must be to completely couple these approaches for practical investigations. At all stages we must continue to employ accurate discretisations and adaptive grids to allow the best numerical solution of the modelled equations. This will allow the use of CFD to be extended in the design process, allowing a shorter and less expensive development cycle for combustion systems coupled with better understanding of their operation.

7.9 Acknowledgements

The author would like to thank Rolls-Royce plc for permission to publish this chapter, as well as colleagues at Rolls-Royce plc, Imperial College of Science, Technology, and Medicine and Loughborough University for many stimulating discussions on modelling of combustion systems.

7.10 References

(1) **Angelberger, C., Veynante, D., Egolfopoulos F.**, and **Poinsot, T.** (1998) Large eddy simulations of combustion instabilities in premixed flames, *Center for Turbulence Research Proceedings of the Summer Program 1998*, pp. 61–82.

(2) **Bicen, A, F., Tse, D. G. N.**, and **Whitelaw, J. H.** (1988) *Combustion characteristics of a model can-type combustor*, Report FS/87/28 (Revised), Department of Mechanical Engineering Fluids Section, Imperial College of Science, Technology and Medicine, London, UK.

(3) **Bond, N. R., Le Vallois J. M.**, and **Menzies, K. R.** (1992) Modelling the vaporiser and primary zone flows for a modern gas turbine combustion chamber, *CFD for Propulsion Applications*, AGARD-CP-510 (AGARD Propulsion and Energetics Panel 77th Symposium).

(4) **Borghi, R.**, (1988) Turbulent combustion modelling, *Prog. Energy Combustion Sci.*, **14**, 245–292.

(5) **Branley, N.** and **Jones, W. P.** (1997) Large eddy simulation of a turbulent non-premixed flame, Eleventh Symposium on *Turbulent Shear Flows*, Grenoble.

(6) **Brocklehurst, H. T., Priddin, C. H.**, and **Moss, J. B.** (1997) Soot predictions within an aero gas turbine combustion chamber, ASME97-GT-148, *International Gas Turbine and Aeroengine Congress and Exhibition*, Orlando.

(7) **Colucci, P. J., Jaberi, F. A., Givi, P.**, and **Pope, S. B.** (1998) Filtered density function for large eddy simulation of turbulent reacting flows, *Phys. Fluids*, **10**, 499–515.

(8) **Coupland, J.** and **Priddin, C. H.** (1985) Modelling the flow and combustion in a production gas turbine combustor, Fifth Symposium on *Turbulent Shear Flows*, Cornell University, pp. 10.1–10.6.

(9) **Gaskell, P. H.** and **Lau, A. C. K.** (1988) Curvature compensated convective transport: smart, a new boundedness-preserving transport algorithm, *Int. J. Numer. Methods Fluids* **8**, 617–635.

(10) **Jones, W. P.** (1980) Models for turbulent flows with variable density and combustion, *Prediction Methods for Turbulent Flows* (Edited by W. Kollmann) Hemisphere, pp. 380–421.

(11) **Jones W. P.** (1994) Turbulence modelling and numerical solution methods for variable density and combusting flows, Chapter 6 of *Turbulent Reactive Flows* (Edited by P. A. Libby and F. A. Williams), Academic Press, pp. 309–374.

(12) **Jones, W. P.** and **Kakhi, M.** (1996) Mathematical modelling of turbulent flames, *Unsteady Combustion* (Edited by F. Culick, M. V. Heitor, and J. H. Whitelaw), Kluwer, pp. 411–491.

(13) **Jones, W. P.** and **Launder, B. E.** (1972) The prediction of laminarisation with a two-equation model of turbulence, *Int. J. Heat Mass Transfer,* **15**, 301–314.

(14) **Jones, W. P.** and **Lindstedt, R. P.** (1988) Global reaction schemes for hydrocarbon combustion, *Combust. Flame,* **73**, 233–249.

(15) **Jones, W. P.** and **Menzies, K. R.** (1997) Calculation of gas turbine combustor flows using an adaptive grid redistribution method, Eleventh Symposium on *Turbulent Shear Flows*, Grenoble.

(16) **Leonard, B. P.** (1979) A stable and accurate convective modelling procedure based on quadratic upstream interpolation, *Comp. Methods Appl. Mech. Engng,* **19**, 59–98.

(17) **Leonard, B. P.** and **Drummond, J. E.** (1995) Why you should not use 'Hybrid', 'Power-Law' or related exponential schemes for convection modelling – there are much better alternatives, *Int. J. Numer. Methods Fluids,* **20**, 421–442.

(18) **Leung, K. M.** (1995) *Kinetic modelling of hydrocarbon flames using detailed and systematically reduced chemistry*, PhD thesis, University of London.

(19) **Maas, U.** and **Pope, S. B.** (1992) Simplifying chemical kinetics: intrinsic low-dimensional manifolds in composition space, *Combust. Flame,* **88**, 239–264

(20) **Manners, A. P.** (1988) *The calculation of the flows in gas turbine combustion systems*, PhD thesis, University of London.

(21) **Pope, S. B.** (1997) Computationally efficient implementation of combustion chemistry using in-situ adaptive tabulation, *Combust. Theory Modelling,* **1**, 41–63.

(22) **Priddin, C. H.** and **Coupland, J.** (1986) Impact of numerical methods on gas turbine combustor design and development, *Calculation of Turbulent Reacting Flows* (Edited by R. M. C. So, J. H. Whitelaw, and H. C. Mongia), ASME, pp. 335–348.

(23) **Shyy, W., Braaten, M. E.,** and **Burrus, D. L.** (1989) Study of three-dimensional gas turbine combustor flows, *Int. J. Heat Mass Transfer* **32**, 1155–1164.

(24) **Spalding, D. B.** (1971) Mixing and chemical reaction in steady confined turbulent flames, 13th International Symposium on *Combustion*, The Combustion Institute, pp. 649–657.

(25) **van Leer, B.** (1974) Towards the ultimate conservative differencing scheme II-monotonicity and conservation combined in a second-order scheme, *J. Computational Phys.* **14**, 361.

(26) **Zhu, J.** (1991) A low diffusion and oscillation-free convection scheme, *Communications Appl. Numer. Methods,* **7**, 225–232.

8

Using CFD to Investigate Combustion in a Cement Manufacturing Process

D Giddings, S J Pickering, K Simmons, and **C N Eastwick**

Synopsis

This chapter presents research into combustion in a cement manufacturing process by The University of Nottingham in association with Blue Circle Cement. CFD models of a cement works precalciner have been developed in which limestone, held in suspension, is calcined to calcium oxide and the endothermic reaction is supported by coal combustion. The work was done using the commercial CFD code Fluent.

The following features of the code were utilized during the study:

- Block structured and fully unstructured meshes;
- Combustion using devolatilization and char combustion models;
- Use of devolatilization models to mimic calcination;
- Particle tracking in a Lagrangian frame of reference with high particle-to-gas mass ratio;
- k–ε turbulence modelling.

Solution difficulties were encountered and solved in the following areas:

- Regions of high field variable gradients;
- Injection of heavily concentrated reacting particles.

Limited validation measurements on the installation indicate fair agreement of the model. The aim of the project was to develop an understanding of the processes within the precalciner to aid in the optimization of the combustion process and the use of alternative fuels. It has been successful in this and has highlighted previously unrecognized features and structures within the precalciner that have implications for future operation.

8.1 Introduction

The aim of this chapter is to present the results of CFD modelling applied to a cement works precalciner. The model includes a full representation of the reactions occurring in the real vessel. More details of this aspect of the work are given in the IMechE Journal of Power and Energy paper (**1**). This chapter describes further work in prediction of tyre chip aerodynamic and combustion behaviour modelling. The precalciner vessel modelled in this work is located at Blue Circle Cement Cauldon works.

8.1.1 The cement manufacturing process

The preheater tower is a counter current heat exchanger approximately 70 m high with typically 4 to 6 cyclone separation vessels. The physical aspects of the preheater tower and the precalciner at the cement works, which were the subject of this study, are illustrated in Fig. 8.1. A large fan that drives the flow through the preheater tower draws hot gases rising from the kiln toward the outlet. Raw meal descends the tower in the opposite direction to the gases. Calcination of the raw meal occurs in the precalciner. Approximately two-thirds of the fuel required for the whole process is delivered to the precalciner in order to support the endothermic calcination reaction and the reaction proceeds in nearly isothermal conditions. The remaining fuel is delivered to the far end of the kiln. Typical configurations of modern plant and a more thorough description of the recent developments are described in the comprehensive reviews by Garret (**2**) and Klotz (**3**).

8.1.2 CFD application

8.1.2.1 Solution settings

Fluent Unstructured (version 4.2.5) was used to develop the model described in the subsequent sections. It was used to solve the steady state, three dimensional, fully viscous Navier–Stokes equations. The k–ε turbulence model of Launder and Spalding (**4**) was used with the constants described in that paper. Reaction mechanisms were simulated using the Magnussen–Hjertager (**5**) eddy mixing model and finite rate devolatilization models using the standard Arrhenius-type rate equation (**6**). The discretization scheme used was PRESTO! (PREssure STaggering Option, see (**7**)) for pressure, as the flow was considered to be subject to high-pressure gradients and body forces. The SIMPLE pressure-velocity-coupling scheme of Patankar (**8**) was used. Six chemical species were included in the model: CO_2, CO, N_2, O_2, H_2O, and lv-vol (the volatile of the coal).

The energy equation was solved using the conservation form in order to improve stability of the calculation in regions of high heat exchange (i.e. where the fastest rates of reactions occur). Simply put in terms of the continuity equation, the forms of the conservation and non-conservation forms are respectively:

$$\frac{\partial \rho}{\partial t}+\frac{\partial(\rho u)}{\partial x}+\frac{\partial(\rho v)}{\partial y}+\frac{\partial(\rho w)}{\partial z}=0 \quad \text{and} \quad \frac{\partial \rho}{\partial t}+\rho\left(\frac{\partial u}{\partial x}+\frac{\partial v}{\partial y}+\frac{\partial w}{\partial z}\right)=0$$

The reason for the effectiveness of this technique is given in the Von Karmen Institute lecture notes (**9**).

8.1.2.2 Grid

For the model of the precalciner, a block structured body fitted grid was developed. The outside surface of the grid is illustrated in Fig. 8.1. One coal inlet is not visible, but it is located directly opposite the coal inlet that is visible on the conical section. It can be seen that the grid in the region of the coal inlet has been refined. The grid refinement was done by the hanging-node method, which divides each hexahedral cell in to eight new hexahedral cells. The cells were refined in the region encompassed by three concentric spheres with coincident centres at the centre of the coal inlet. Thus, three refinement regions can be seen on the

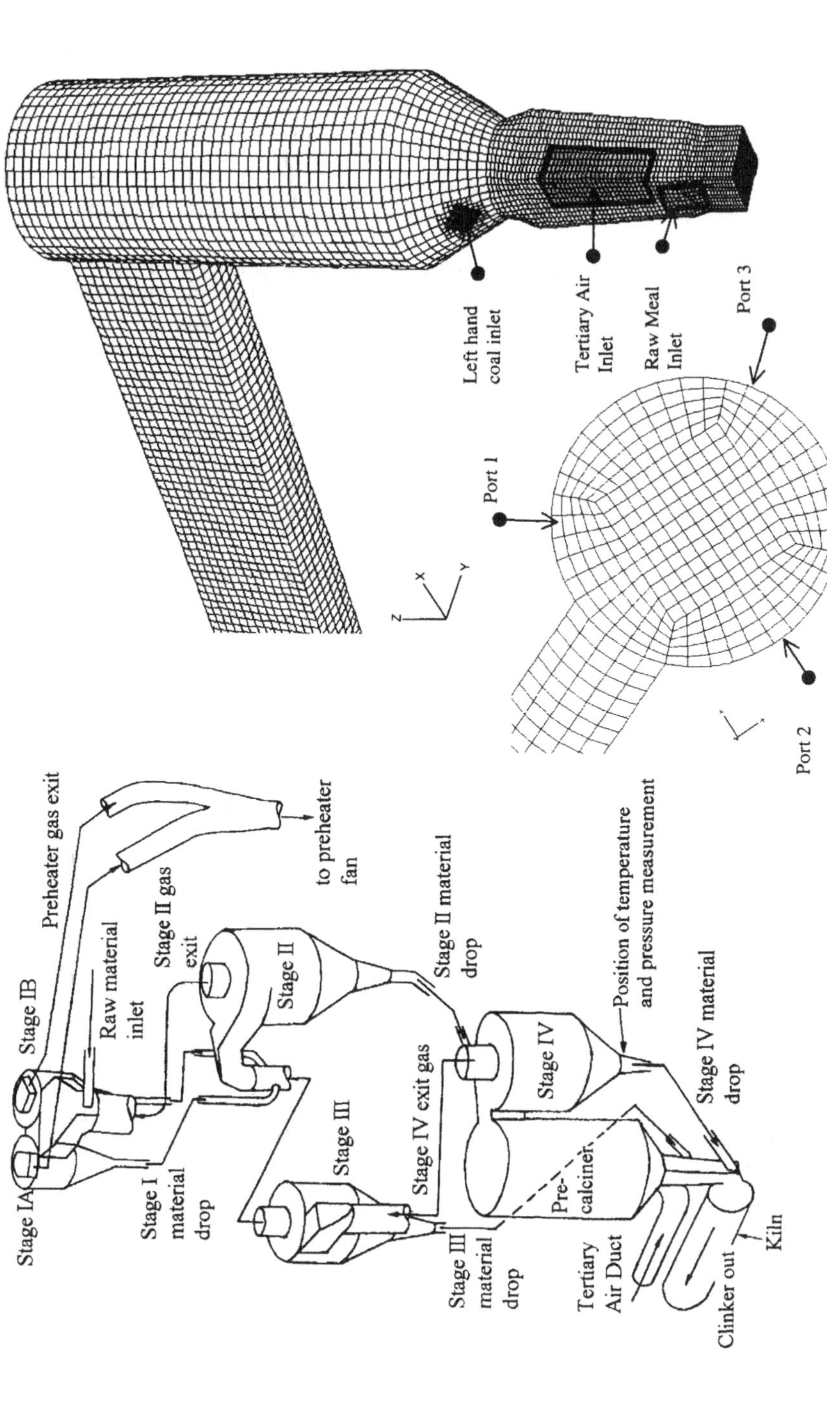

Fig. 8.1 The preheater tower and precalciner model

surface of the model. The outlet duct was added to allow the flow to develop before the exit. The total number of computational cells was 45 000. The height of the model was 25.84 m and the diameter of the main cylindrical section was 6.9 m.

8.1.2.3 Solid particle modelling

The three particulate types were modelled using the Lagrangian frame of reference (7). They represented coal, raw meal, and tyre chips. A coupled solution was used, in which interaction of the discrete phase (the particles) with the continuous phase (the fluid) is considered.

The particles in the real precalciner are driven upwards against gravity by the gas velocity. The mass injection rate of raw meal particles to the model was 55 kg/s, the coal injection rate was 1.5 kg/s, and tyre chips 0.5 kg/s. The heavy raw meal load caused solution stability problems. It was found that increasing the number of raw meal particle tracks, spreading the initial injection, and employing stochastic effects helped stability. Stochastic dispersal of the particles was employed (7). Each stochastic calculation makes random changes to the turbulent component of the velocity vector, u', in each computational cell to alter the trajectory of the particle track. The total mass for each injection is then divided equally between the number of stochastic calculations, 20 in this case. The Rossin–Rammler size distribution was applied to the coal and raw meal particles (Table 8.1).

Table 8.1 Rossin–Rammler size distribution of coal and raw meal particles

Minimum particle diameter	20μm
Mean particle diameter	55μm
Maximum particle diameter	100μm
Rosin–Rammler spread parameter	1.04
Number of stochastic attempts	20
Interval length of tracking steps	10 mm
Number of tracking steps	15,000

The initial size of the tyre chips was determined from aerodynamic experimental work in a wind tunnel. A swelling coefficient (7) was chosen to cause all tyre chips to have the same aerodynamic characteristics as a 3 mm-diameter piece of char with density 70 kg/m^3 when the volatiles have completely yielded. This behaviour was inferred from experimental combustion of tyre chips in an ashing furnace at 900 °C. Tyre chips decomposed to char granules of this size once the volatile component had been released. The tyre chip data is presented in the following Table 8.2.

Table 8.2 Tyre chip injection data

Chip No.	% injection	Initial dia (cm)	Swelling coefft	Tyre chip mass (g)
1	4	8.8	2.38	8.2
2	4	6.1	3.43	8.2
3	15	8.9	3.73	20.6
4	15	6.5	5.08	20.6
5	16	8.1	4.81	28
6	16	6.1	6.26	28
7	8	9.3	4.56	34
8	8	6.3	6.7	34
9	7	13.7	3.92	54
10	7	6.7	8.03	54

The devolatililization model of the combusting particle type was used to simulate the calcination reaction. The single rate model was used of the Arrhenius form, $k = A\exp(-E/RT)$. The activation energy (E = 2.05×10^8 J/kmol) and the pre-exponential constant (A = 3.81×10^8 s^{-1}) were determined by reference to the work of Borgwardt (**10**).

Coal and tyre chip combustion was included in the model. The rate of devolatilization of coal was set according to the model of Badzioch and Hawkesley (**6**). The devolatilization rate of tyre material was set according to experimental data; a constant rate was used, which caused all volatiles to yield in 30 seconds. Combustion of carbon (char) was considered as a two step reaction (**11**), first to CO, then to CO_2. The diffusion-limited rate was used, whereby the diffusion rate of the gaseous oxidant to the surface of the particle determines the oxidation rate (**12**). Data for the composition of a medium volatile coal was obtained from Smoot (**13**). The coal simulated was based on Upper Freeport MVD. Tyre composition was determined from other research work (**14**). The compositions are stated in Table 8.3. Proximate analysis and ultimate analysis are defined by Borman and Ragland (**11**).

Table 8.3 Composition of model coal and tyre chip material

		Model coal	Model tyre
Proximate analysis (by mass, %)	volatiles	18.6	62.1
	char	74.4	26.1
	ash	7	11.8
Ultimate analysis (by mass, %)	carbon	87.5	85.3
	hydrogen	4.8	11
	oxygen	7.7	0
Calorific value (kJ/kg)		30 985	36 924

8.2 Results

Figure 8.2 shows some features of the model precalciner operation. The outline shows the edges of the precalciner vessel, with the exit duct directed away from the point of observation. Velocity vectors are projected from lines across the centre of the vessel at 10 heights. Temperature contours are plotted on two planes, at 11.4 m, and at 20.4 m from the base. These are at the height of available access ports in the wall of the real precalciner, which will be referred to in the validation section. The behaviour of the particles in the model are represented by three coal particle trajectories from each of the coal inlets and six tyre chip trajectories from the tyre inlet. Some qualitative aspects of the predicted behaviour will be drawn from this figure in the following sections.

8.2.1 CFD results

The model shows that particles injected at the raw meal inlet (RMI), are strongly influenced by the gases injected at the tertiary air duct (TAD). Of the mass of raw meal injected approximately 85 per cent is driven across to the far side. The momentum of the gases from the TAD is 27 kg/s at 30 m/s compared to 55 kg/s of raw meal particles at approximately 15 m/s. The effect of turbulence in the real installation tends to disperse the particles in random patterns. CFD shows that the initial common trajectory is toward the right-hand side of the precalciner in the conical section, despite turbulence effects. The effect of turbulence is significant in dispersing the particles following their passage beyond the throat of the conical section. There is then significant recirculation of some particles in the main cylindrical section before exit. The time a particle will take to traverse from inlet to exit thus varies considerably;

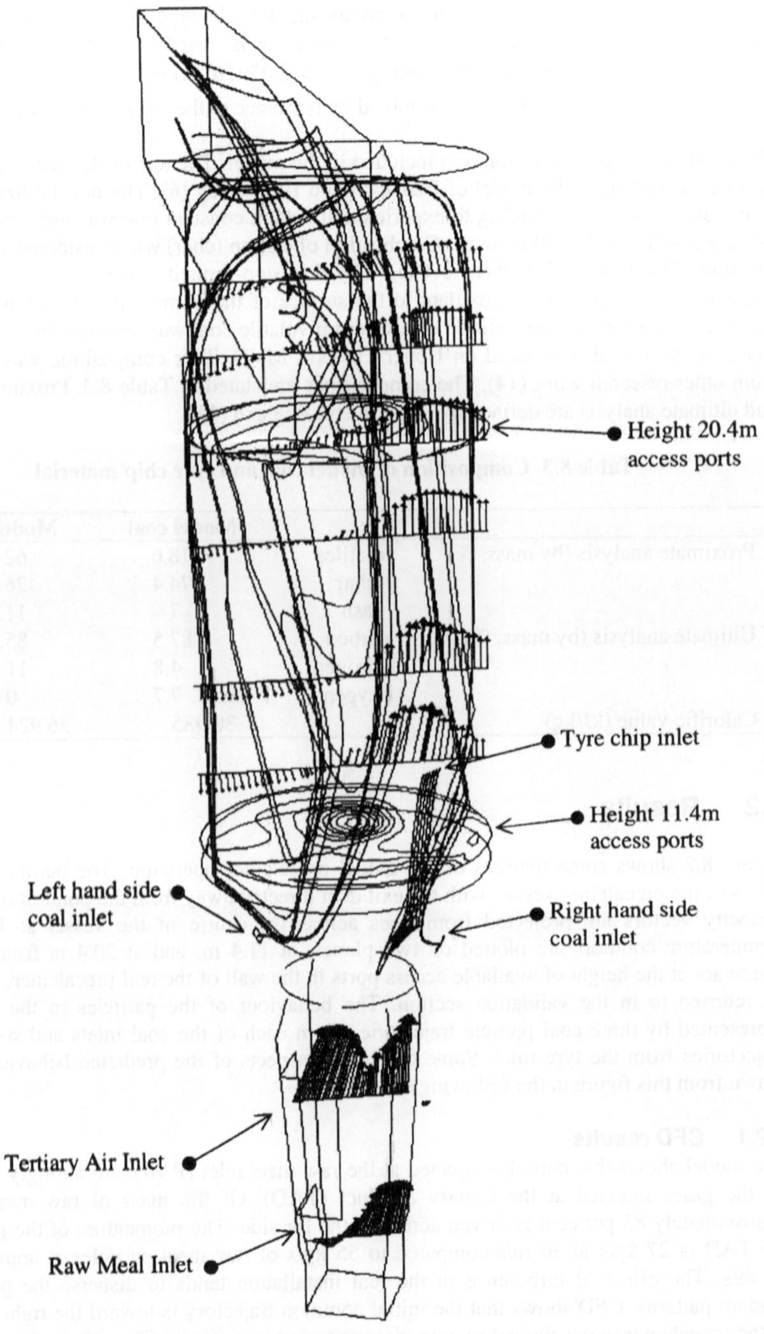

Fig. 8.2 Features of particle trajectories, velocity vectors, and temperature contours

an average of 10 seconds is apparent from the model, with some taking as little as 5 seconds. The momentum of the gas mixture drives the particles. Therefore, the model flow field, illustrated in Fig. 8.3 by velocity vectors, shows why the particles are observed to behave as they do. Recirculation of the gas on this plane is strong. On the left-hand side of the precalciner, the downward velocity reaches 11 m/s and on the right-hand side the upward velocity reaches 20 m/s. An interesting feature shown in the velocity vectors is in the region of the raw meal inlet where the high injection rate of raw meal causes such a depression in the upward velocity that the flow is forced slightly downwards at one position.

The model suggests that in the order of 85 per cent of the injected raw meal has undergone the calcination reaction fully before exit from the precalciner. No measurement on the real installation has yet been performed that can verify this. The post-processing available shows the discrete phase devolatilization rate per cell from the coal and the raw meal in one field variable. This includes the volatiles (lv-vol) from the coal and the CO_2 from the raw meal, which is modelled as a volatile despite the true nature of the reaction. The model suggests that calcination occurs mainly toward the wall at the right hand side of the precalciner, i.e. on the side furthest from the TAD above the conical section. Since calcination is a highly endothermic reaction, an associated region of relatively low temperature would be consistent with a high rate of calcination. This is what is observed in the model; the region corresponding to high calcination rate in the foregoing discussion shows a depression to the range 900 °C to 1000 °C. Energy supplied by the coal combustion reactions maintains the temperature until the calcination reaction is complete.

The behaviour of the coal in the model is consistent with the above discussion. Coal from the inlet on the same side as the Tertiary Air Duct (left-hand side coal inlet) rises and exits without recirculating. Coal from the other inlet (right-hand side coal inlet) rises to the ceiling of the precalciner and recirculates in the full length of the main cylindrical section. Residence times are 0.6 to 1.0 second for coal from the left-hand side and 5 to 18 seconds for coal injected from the right-hand side. Volatiles are released very quickly due to the high temperature of the gases into which the relatively cold (75 °C) coal air stream enters, and the conversion to CO is completed before the exit from the main body of the precalciner. Char oxidation is completed before exit from the precalciner for Coal B; 20 per cent of the char from Coal A is not oxidized before exit. Some CO is present at the exit from the precalciner at 0.46 ppm, confirming the presence of reacting char.

Tyre chips fall and then recirculate similar to the left-hand side injected coal. Residence time in the precalciner is sustained by the recirculation.

8.2.2 Validation data from the vessel

Six access ports are present on the cylindrical section of the precalciner. Three are equally spaced on the circumference at two heights (Fig. 8.1); 0.6 m above the conical section (elevation 11.4 m on the model) and 5 m below the top (elevation 20.4 m). The data from the model at these two heights is presented in Fig. 8.3. Contours of velocity in the vertical, z direction (m/s), temperature (°C), and particle concentration (kg/m^3) were related to measurements made on the precalciner. Probes were inserted approximately 0.3 m into the flow from the edge of the refractory lining through the access ports.

All temperature measurements show good correlation to within 60 °C.

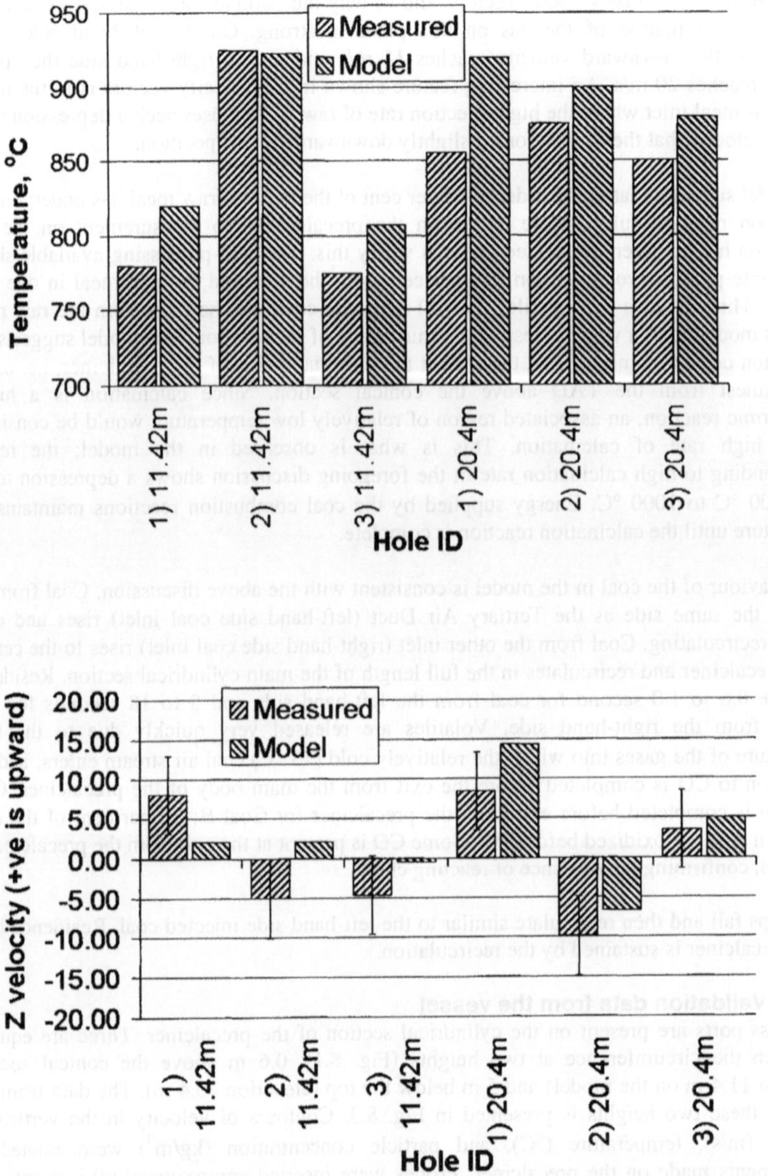

Fig. 8.3 Comparison of site measurements and CFD results for temperature and velocity with error bars applied to the velocity measurements

Accurate measurement of gas velocity inside the precalciner is very difficult on account of the very high particulate loading. However, for the purposes of initial validation, an S-type Pitot probe was used to give and indication of flow direction in the precalciner.

Figure 8.3 shows that there is considerable uncertainty in the actual velocity measurements. The relative magnitude and flow direction at each measurement port shows similarity to the measurements. Error bars on the measured values are used to give an estimate of the uncertainty. The probe blocked up with particles after a few seconds and had to be cleared by intermittent purging with air. There was significant unsteadiness in the velocity measurements of the probe. Nevertheless, it can be seen from Fig. 8.4 that there is generally good agreement between the prediction of the flow velocity and the direction and relative magnitude of the velocity indicated by the measurements.

8.3 Conclusions

The behaviour of particles within the precalciner is of interest to Blue Circle. The data available from the model can be used to make decisions about best injection technique for coal and tyre chips. The model shows that the points of injection give good combustion features because of the recirculation characteristics leading to prolongued residence times.

The work-to-date has illustrated some important characteristics of the behaviour in the precalciner vessel, which cannot be practically measured. The usefulness of the CFD tool for this work has been proven in its ability to make clearer the behaviour of the vessel. Validation of results can be done by making measurements at key positions through the precalciner wall to verify the information produced by the model at those locations. Sensitivity analysis can also be done on the model by changing some of the parameters.

8.4 Acknowledgements

The authors gratefully acknowledge the financial and practical support from Blue Circle Cement (Cauldon Business Unit) given to undertake this research project. Support from EPSRC is also gratefully acknowledged.

8.5 References

(1) **Giddings, D., Eastwick, C. N., Pickering, S. J.,** and **Simmons, K.** Computational fluid dynamics applied to a cement precalciner, IMechE *J. Pwr Energy*, Accepted for publication 23/6/99.

(2) **Garret, H. M.** (1985) *Precalciners today – a review*, Rock Products, July.

(3) **Klotz, B.** (1997) New Developments in Precalciners and Preheaters. 1997 *IEEE Cement Industry Conference*, Hershey, Pennsylvania, 21–24 April.

(4) **Launder, B. E.** and **Spalding, D. B.** (1974) The Numerical Computation of Turbulent Flows, *Comp. Methods Appl. Mech. Engng*, **3**, 269–289.

(5) **Magnussen, B. F.** and **Hjerteger, B. H.** (1976) On mathematical modelling of turbulent combustion with special emphasis on soot formation and combustion, Sixteenth International Symposium on *Combustion*, The Combustion Institute, pp. 719–729.

(6) **Badzioch, S.** and **Hawkesley, P. G. W.** (1970) Kinetics of thermal decomposition of pulverised coal particles, *Ind. Engng Chem. Process Des. Develop*, **9**, 521–530.

(7) Fluent Incorporated. User's Guide for FLUENT/UNS and RAMPANT, Release 4.0, April 1996, Vol. 1, 2, and 4, Fluent Incorporated, Lebanon, NH 03766, USA.

(8) **Patankar, S. V.** (1980) *Numerical Heat Transfer and Fluid Flow*, Hemisphere, Washington, D.C.

(9) **Anderson, J. D.** (1990) *Introduction to Computational Fluid Dynamics*, Chapter 2, Von Karmen Institute for Fluid Dynamics, Belgium.

(10) **Borgwardt, R. H.** (1985) Calcination kinetics and surface area of dispersed limestone particles, *AIChE Journal*, **31**, 103–111.

(11) **Borman, G. L.** and **Ragland, K. W.** (1998) *Combustion Engineering*, McGraw-Hill, USA.

(12) **Baum, R. K.** and **Street, J. H.** (1971) Predicting the combustion behaviour of coal particles, *Combust. Sci. Tech.*, **3**, 231–243.

(13) **Smoot, L. D.** (Editor) (1993) *Fundamentals of Coal Combustion*, Elsevier, Amsterdam, London.

(14) **Conesa, J. A., Font, R., Fullana, A.,** and **Caballero, J. A.** (1998) Kinetic modelling for the combustion of tyre wastes, *Fuel*, **77**, 1469–1475.

9

Validation of the Coal Combustion Capability in the Star-CD Code

A Ghobadian, F Lee, and **P Stephenson**

9.1 Introduction

When any major enhancement is made to a commercial CFD code, both the code developer and the code user have to assure themselves that the code is performing correctly. In addition, the user has to make sure that he or she correctly understands the code enhancement and how to use it. This chapter illustrates this process by describing the validation of the coal combustion capability to the established Star-CD program, developed by Computational Dynamics (CD). The approaches used by the code developer and code user are described and contrasted.

Coal combustion provides a challenging example of code validation for a number of reasons (see, for example, references (**1**) and (**2**). The physical processes being modelled are not fully understood and are very complex. The submodels used in CFD codes generally have to simplify these processes and still require substantial computing resources. Also, these submodels require values of coal-dependent properties which are difficult to obtain. Finally, few measurements are available for full-scale utility furnaces as such measurements are both technically difficult and very expensive.

9.2 Modelling coal combustion using a CFD code

9.2.1 Why model coal combustion?

Pulverized coal is the largest fuel source for electricity generation, and is likely to account for 70 per cent of new capacity. The successful application of CFD to coal combustion can assist plant suppliers in the design of new plant or improvements to existing plant. It can also assist plant operators to find safe, efficient ways of operating plant with minimum environmental impact, and to assist in fuel selection.

9.2.2 The coal combustion model in Star-CD

For the work reported here, Star-CD has been used with non-orthogonal unstructured hexahedral meshes. Turbulence is modelled using the k–ε model and second order differencing has been used in some predictions. Gaseous radiation is modelled using the discrete transfer method (**3**). A Lagrangian method is used to track the particles. Devolatilization is calculated using the single step model of Badzioch and Hawksley (**4**) and char combustion is calculated using the model of Field, Gill, Morgan, and Hawksley (**5**). Gas phase combustion is modelled by a variant of Spalding's conserved scalars approach. A single step reaction is used; the code solves for mixture fractions for evolved volatiles and amount of char consumed and the actual composition is found by using a mixed-is-burnt assumption. The NOx modelling includes thermal, fuel, and prompt NOx. Thermal NOx is calculated from an extended Zeldovich approach and fuel and prompt NOx are calculated from the method of De Soete (**6**).

9.3 Validation by the code developer

CD's aim in the validation was to be able to demonstrate to potential users both that their code gave predictions that were in reasonable agreement with measurements, and that it could be successfully applied to what a user would want to model, namely a full-scale utility furnace. As detailed measurements are rarely available for a utility furnace, CD's validation was made in two parts. The first involved modelling a full-scale single burner test rig, for which detailed measurements were available, and the second involved modelling a power station furnace, for which only limited measurements were available. These two validation exercises will now be outlined.

9.3.1 Modelling a full-scale single burner test rig

Testing of a single full-scale low NO_x burner was undertaken on Mitsui Babcock's 40 MW_t large scale test facility, as part of the UK Department of Trade and Industry (DTI) collaborative project on coal quality and NO_x (**1**). The burner fires horizontally into a water-cooled furnace of 4.2 by 4.2 m cross-section and 15 m length. The walls are partially covered with refractory to reproduce the temperatures expected in a utility furnace. Tests were conducted using South Brandon coal. The facility has recently been replaced by a new test rig of up to 90 MW_t capacity. A diagram of the 40 MW_t rig is shown in Fig. 9.1.

Mitsui Babcock provided measurements of temperature and species concentrations on horizontal traverses for a number of measurement ports, and velocity measurements were obtained by BG plc. These detailed measurements provided a valuable check on the computer predictions. A typical comparison is shown in Fig. 9.2.

It is not reasonable to expect very close agreement between prediction and measurement for this situation, as the physical processes are complex and not fully understood and the measurements themselves are made in a hostile environment. The code user has to decide the level of agreement that it is reasonable to expect; in this case, that expectation was met. Predicted exit values of NO_x and unburned carbon loss are compared with the Mitsui Babcock measurements in Table 9.1.

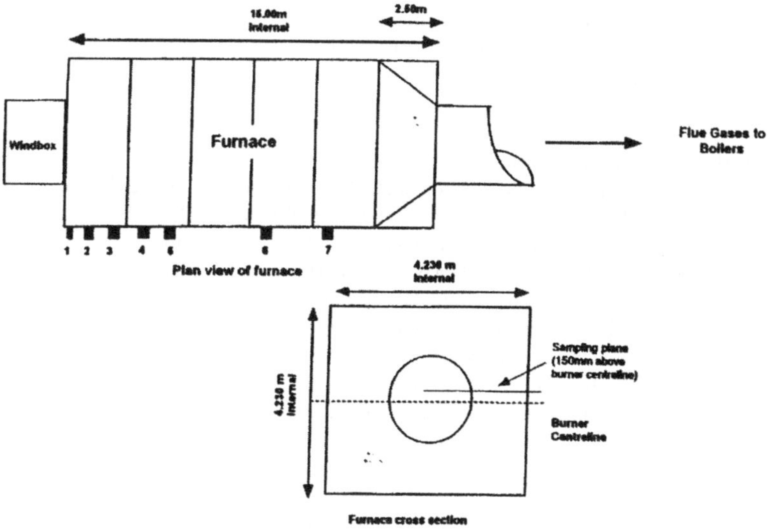

Fig. 9.1 Diagram of Mitsui Babcock 40 MW rig, showing measurement ports

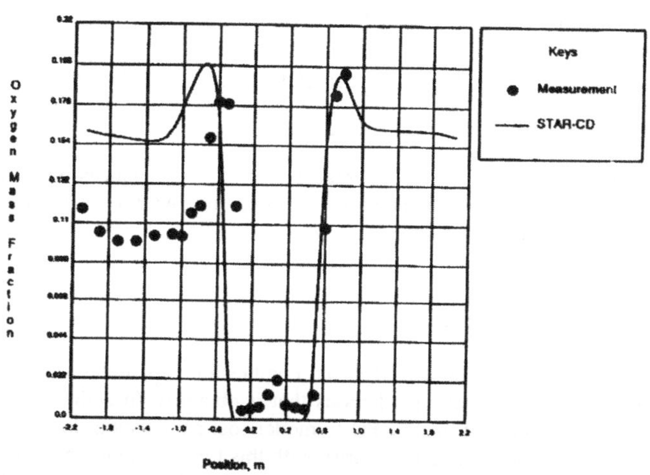

Fig. 9.2 Comparison of oxygen concentrations at port 2

Table 9.1 Comparison of predicted and measured NO_x and unburned carbon loss for Mitsui Babcock 40 MW rig, burning South Brandon

	NO_x, mg/Nm^3, dry at 6% O_2	Unburned loss (% GCV)
Measurement, Mitsui Babcock	559 (a), 625 (b)	0.5
Star-CD prediction	636	3.2

Note: (a) value for modelled test, (b) value averaged over several tests

The agreement between measured and predicted NO_x values is reasonable; the predicted unburnt loss is less satisfactory. This is in line with the overall conclusions from all the predictions in this DTI project, namely that agreement with NO_x was generally encouraging but that burnout predictions were more difficult.

9.3.2 Modelling a power station furnace

Combustion testing was also undertaken on one of the furnaces at Innogy's Didcot power station as part of the DTI collaborative project on coal quality and NO_x. This furnace is front-wall fired. It has a total of 48 burners arranged in 4 tiers of 12 burners per row. Coal is pulverised by 8 mills, and full load (500 MW electrical) can be maintained with 36 of the 48 burners firing. Tests were conducted using Pittsburgh#8 coal. Didcot is of Mitsui Babcock design and has their Mk 3 low NO_x burners installed. Further tests were conducted at Powergen's Ratcliffe power station, which is of the same construction as Didcot. Mitsui Babcock carried out a performance analysis to provide boundary conditions (eg coal and air flows and heat to walls and superheaters) to provide boundary conditions for the CFD predictions.

CD's computer model of a Didcot furnace assumed symmetry and, therefore, included half the furnace, with 24 burners, of which 16 were firing. The only measurements were the exit values of NO_x and unburned carbon loss, and these are compared with predictions in Table 9.2.

Table 9.2 Comparison of predicted and measured NO_x and unburned carbon loss for Innogy's Didcot furnace, burning Pittsburgh#8

	NO_x, mg/Nm^3, dry at 6% O_2	Unburned loss (% GCV)
Measurement, Innogy	627	1.65
Star-CD prediction	691	1.75

The agreement between measured and predicted NO_x values is again good. The predicted unburnt loss is also in good agreement with measurement, but this is partly because of adjustments made to the char kinetic data.

9.4 Validation by the user

Innogy, as user, also wanted to validate the coal combustion capability within Star-CD, so that it could have confidence in any predictions that it obtained. In addition, there was no prior experience within Innogy of using CFD to model coal combustion. Innogy's validation therefore, had also to include gaining familiarity with this type of modelling, both in general terms and specifically in how to model it using Star-CD. They therefore started with a simpler case, namely a Drop-Tube Furnace, and are currently modelling their own single burner Combustion Test Facility.

9.4.1 Modelling a Drop-Tube furnace

A Drop-Tube Furnace (DTF) is an experimental facility used to measure coal devolatilization and combustion rates (see Fig. 9.3).

The CRE DTF is a vertical furnace of length 2 m. Pulverized coal is fed (typically at 15 g/hr) into the top of the furnace in a carrier gas. Most of the gas enters the furnace via an annular region around the feeder tube. The gas consists of a mixture of nitrogen and oxygen, with varying amounts of oxygen. The coal mixes with the preheated gas and devolatilizes. Volatile gases are released, leaving char and ash. If sufficient oxygen is present, combustion will occur. Any residue is collected at a feeder tube at the bottom of the drop tube. The furnace is surrounded by electrical heaters and thermal insulation to maintain the furnace walls at the required temperature.

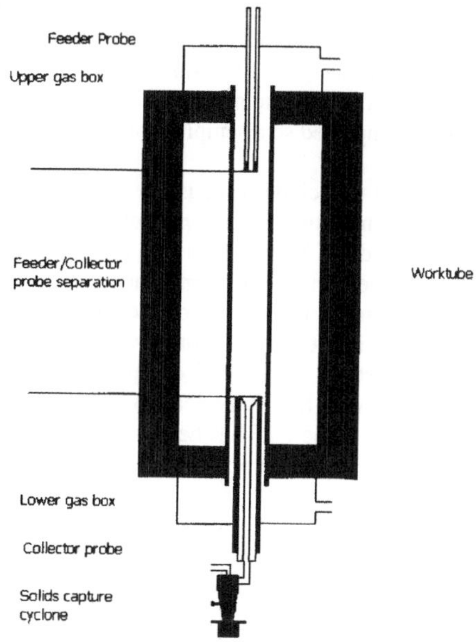

Fig. 9.3 Diagram of CRE's Drop-Tube furnace

CRE provided non-reacting gas tracer measurements and Innogy's CFD model was checked against them, to ensure that the flow field was being correctly predicted. Predictions were then made with coal combustion. One problem was that the predictions are very dependent on char kinetic parameters and these, in turn, are often derived from the DTF tests. Therefore, comparisons were made with CRE measurements for Thoresby that had been made as part of the DTI NO_x project, and the char kinetic data were taken from an independent analysis of these measurements. A typical prediction between measured and predicted burnout (defined as the per cent weight loss on a dry ash free basis) is given in Fig. 9.4.

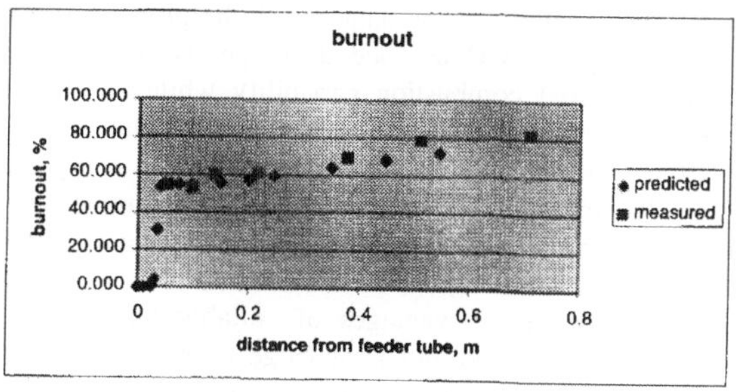

Fig. 9.4 Typical comparison between predicted and measured burnout

This modelling provided Innogy with a very useful introduction to CFD and coal combustion, and the agreement with measurement was satisfactory.

9.4.2 Modelling a single burner test furnace

Innogy is currently modelling its own single burner Combustion Test Facility (CTF). This has a maximum thermal input of 0.5 MW and a combustion chamber 0.8 by 0.8 m in cross-section and 4 m long. As a first step, isothermal predictions were obtained for both a scaled version of the Mitsui Babcock low-NO_x burners (as in Section 9.3) and a scaled version of a completely different type of burner, designed for semi-anthracite coal. In the latter case, some velocity measurements were available and the predictions were in reasonable agreement both with these measurements and a simplified semi-empirical model of the main burner flow.

This work has now been extended to include combustion and radiation for both burner types; the Mitsui Babcock low NO_x burner studies form part of Innogy's contribution to the DTI collaborative project on coal blending. Much effort has been required to obtain predictions that exhibit the main features of the flames (e.g., firmly attached to the burner and occupying only the central portion of the furnace, in the case of the Mitsui Babcock low NO_x burner). This has now been achieved, and, in the process, Innogy has gained much useful experience in applying CFD to coal combustion. A detailed comparison between measurements and predictions is now in progress and will, hopefully, lead to further code validation. In addition, the CFD model of coal combustion in the CTF has been used in a collaborative research effort on the effects of coal blending on combustion performance (7).

9.5 Discussion

An ideal validation would involve comparing predictions with detailed measurements for well-controlled tests on the plant that the user wishes to model. Unfortunately, it is very difficult, if not impossible, to achieve this for coal combustion. Controlled tests on actual utility furnaces are very expensive, the exact test conditions cannot always be determined, and only limited measurements can be made. More detailed measurements can be obtained on test rigs, but they still provide a challenging environment. The physical processes occurring in coal combustion are complex and not fully understood, and the sub-models used in most CFD codes have to simplify matters. These sub-models require values for coal-dependent properties that can be derived only from specialist laboratory-scale tests. Also, the predictions are often very sensitive to the detailed inlet conditions, but these are not always known in sufficient detail.

As a result of these points, any validation has to make the best use of whatever data are available, and allow for shortcomings in the data and in the physical sub-models. From this viewpoint, the validation made by CD as code developer is acceptable; it gives the user confidence to try using the coal combustion capability while, at the same, showing its shortcomings (for example, in the prediction of carbon burnout). Innogy's validation, as code user, has been very successful as a way of gaining experience in using the Star-CD code to model coal combustion. Its first phase (modelling a drop tube furnace) has been successful and encouraging progress has been made with the second stage (the single burner furnace).

All the validations have shown the advantages of collaborative projects, where suitable experimental data sets can be obtained and knowledge shared with other CFD modellers. They also show the need for further experimental data and for improved coal sub-models. In view of the computing requirements, parallel computing and improved accurate and flexible meshes need to be investigated.

9.6 Conclusions

The process of validating a major new feature in a CFD code has been illustrated by using the coal combustion model in Star-CD as an example. The different requirements of the code developer and user have been described. The validation is inevitably limited by the complexities of the physical processes being modelled and the shortage of detailed measurements for actual plant. The advantages of collaborative projects, such as the DTI projects on NO_x and on coal blending, have been demonstrated.

9.7 Acknowledgements

The authors gratefully acknowledge the sources of the measurements used in this chapter. The 40 MW_t single-burner test-rig results were obtained by Mitsui Babcock and BG plc. The Drop-Tube Furnace results were obtained by CRE Group Plc, and the Didcot Power Station measurements were provided by Innogy plc.

Some of the results reported here were obtained as part of the recent DTI collaborative project on Coal Quality, NO_x, and carbon burnout. The authors wish to acknowledge substantial UK Government support for this work. The total cost of the project was approximately £2.1 million, with £742 000 being contributed by the Department of Trade and Industry's Clean Coal Technology Programme and £668 000 by the Engineering and Physical Sciences Research Council, the latter specifically for the university activities. Some of the other results reported here were supported by a current collaboration on coal blending and combustion, and this is also receiving funding via the DTI.

9.8 References

(1) **O'Connor, M.** (1999) The effects of coal quality on NOx emissions and carbon burnout in pulverised coal-fired utility burners, ETSU report COAL R153.

(2) **Eaton, A. M., Smoot, L. D., Hill, S. C.,** and **Eatough, C. N.** (1999) Components, formulations, solutions evaluation and application of comprehensive combustion models, *Prog. Energy Comb. Sci.*, **25**, p. 387.

(3) **Lockwood, F. C.** and **Shah, N. G.** (1981) A new radiation solution method for incorporation in general combustion prediction procedures, Eighteenth International Symposium on *Combustion*, pp. 1405–1414.

(4) **Badzioch, S.** and **Hawksley, P. G. W.** (1970) Kinetics of thermal decomposition of pulverized coal particles, *Ind. Engng Chem. Process Des. Develop.*, **9**, 521–530.

(5) **Field, M. A., Gill, D. W., Morgan, B. B.,** and **Hawksley, P. G. W.** (1967) The combustion of pulverised coal, BCURA.

(6) **De Soete, G. G.** (1975) Overall reaction rates of NO and NO2 formation from fuel nitrogen, Fifteenth International Symposium on *Combustion*, The Combustion Institute, pp. 1093–1102.

(7) **Beeley, T., Cahill, P., Riley, G., Stephenson, P. L., Lewitt, M.,** and **Whitehouse, M.** (2000) The effect of coal blending on combustion performance, Report no. Coal R177, DTI/Pub URN 001509, March.

10

Blast Wave Simulation

M Docton, S Rees, and S Harrison

Synopsis

Frazer-Nash Consultancy has developed and validated a methodology to predict fluid behaviour subsequent to a blast using CFD. Traditional analysis in this area relies on specific testing or, more typically, scaling standard blast data. The use of CFD allows the transient pressures developed on nearby structures to be readily predicted. Analysis of the subsequent response and deformation of these structures, including human injury aspects, may then be undertaken using these CFD predictions.

Typical blast scenarios include accidental pipe rupture, explosions within buildings, and vehicles encountering mines. The use of CFD for blast simulation facilitates the design of blast resistant structures/surfaces and the prediction and mitigation of human injury.

Examples of blast simulations are presented, including validation, together with typical examples of subsequent structural analyses.

10.1 Introduction

The design of blast-resistant structures is typically undertaken by a combination of engineering experience, design guidelines, and trials. Examples of typical blast simulation applications are:

- Vehicles encountering mines;
- Rupture of pressurised piping;
- Explosions near to buildings.

Where the structure being designed is similar to an existing system and the blast source is well understood, experience and the use of design guidelines can be a very cost-effective design methodology. However, where the blast source and/or structure is less well understood a 'rule of thumb' design and trials process can prove to be very costly. The problems of the traditional methods for design are compounded as requirements for human safety and structural integrity become more stringent.

This chapter provides an outline to an alternative design process for blast mitigation, which utilizes a combination of design experience, empirical data and computer modelling. The methodology is based around three aspects:

- Prediction of the blast and the developed pressures on nearby structures;
- Subsequent kinematic response and deformation of the structure;
- Response of any occupants associated with the structure.

This chapter concentrates on the primary aspect, namely the prediction of the blast, however, typical examples of subsequent analyses are presented. The blast modelling methodology developed and validated by Frazer-Nash Consultancy Limited (FNC) has been based on a variety of simple test-cases, for example, normal and angled shock reflection. The predictions have been compared to empirical and analytical results and subsequent blast simulations have been undertaken on more complex structures.

The basis of the CFD modelling involves the conversion of the blast source into a transient input within the model. This can be achieved in one of two ways; namely a pressure/temperature source or a burn model. This transient input can be correlated to a known blast magnitude. Once the blast has been enabled the developed pressures in the flow domain can be monitored and the subsequent response of a given structure predicted.

10.2 Blast modelling

10.2.1 Overview

Before blast modelling can be undertaken it is important to understand the phenomena associated with a typical blast. When an explosive detonates it rapidly burns and generates a ball of explosive products at an extremely high pressure and temperature. This ball of high pressure then expands and compresses the air immediately surrounding the explosion, which in turn expands and compresses neighbouring regions. Thus a pressure wave begins to propagate away from the explosion. The high pressure of the air in the pressure wave means that its temperature and sonic velocity are increased. This pressure wave can, therefore, travel at high speeds, for example Mach 8. As the pressure wave expands it will continue to propagate at high speed and eventually detach itself from the expanding ball of explosion products, whose expansion rapidly slows as it expands. In addition, the pressure wave will quickly form into a shock wave as the gas high pressure 'catches up' with the lower pressure gas in front of it.

One further effect should be mentioned before proceeding. The expanding explosion products are at high temperature and can combust causing further energy to be released. However, this combustion effect is only important in a limited number of explosive events.

In summary the result of an explosion is a blast which has two components: a *blast wind* and a *shock wave*. The former comprises the slowly expanding ball of explosion products (which may be combusting) and the air which is entrained/pushed in front of it. The latter is the detached shock. Both these elements of the blast can be important in causing damage to nearby structures. However, as the distance from the blast increases the shock wave becomes more damaging, as it can maintain a high pressure over a greater distance.

This study is primarily concerned with modelling the shock wave, which causes significant damage over greater distances than the blast wind. However, the blast wind is implicitly modelled using the methodology adopted.

10.2.2 Methodology validation

The validation exercise was divided into two areas: firstly an examination of the predicted behaviour of shock wave reflection at various incidences and secondly an examination of the decay of a shock wave in free space. The validation test cases included the following:

- Two-dimensional shock wave wall reflection, at normal and 20 degrees incidence;
- Three-dimensional shock wave propagation in free space.

10.2.2.1 Wall reflection

For the wall reflection validation the parameters examined were the overpressure received by the wall from the incident shock wave and the predicted angle of the reflected blast. The models used for this were unsteady, compressible, and inviscid using quadrilateral meshes. A transient blast pulse was introduced at one side of the domain and allowed to propagate towards a wall at the other side. This pressure pulse may be created by a variety of methods within the CFD model. For example, a discrete fluid zone may be initialized with a high pressure and temperature or alternatively a burn model may be used. In this instance, a time dependent pressure and temperature boundary condition was utilized. In any approach it is essential to relate the desired blast magnitude to an equivalent initial pressure and temperature, and it is in this area that FNC have developed the necessary methodology.

The computational domain was bounded by symmetry planes. Figures 10.1 and 10.2 illustrate the computational domain and pressure contours of an incoming shock wave respectively.

When a shock wave impacts on a surface the local particle velocity is reduced to zero, then reflected at with an equal and opposite velocity. This initial reflected part of the wave then moves back toward the source through the oncoming remainder of the shock wave. Thus although the reflected and unreflected wave have equal and opposite particle velocities, their characteristics are different as they have travelled through different (each other's) media. The resulting reflected pressure can be up to eight times the incident pressure.

The pressures developed by the shock wave on the wall were recorded, along with the pressure decay in the wake of the reflected wave. The predicted pressures in the domain are shown by Figs 10.3(a) and 10.3(b), for before and after reflection respectively.

Fig. 10.1 Domain for two-dimensional normal reflection

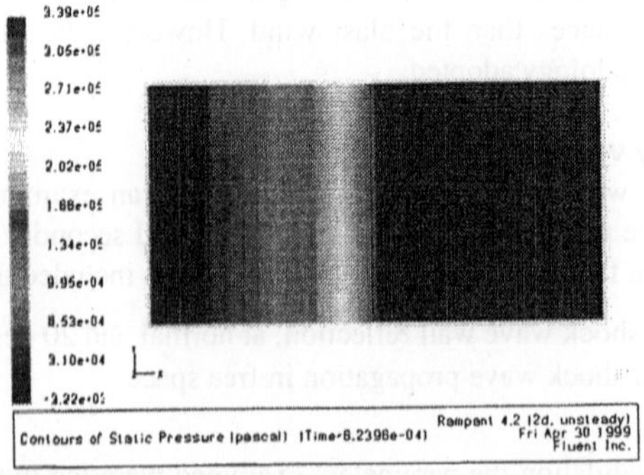

Fig. 10.2 Pressure contours for an incoming shock wave

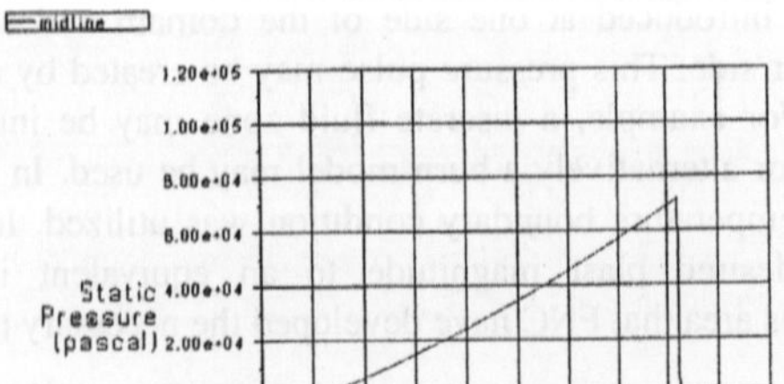

Fig. 10.3(a) Pressure predictions across the domain before reflection

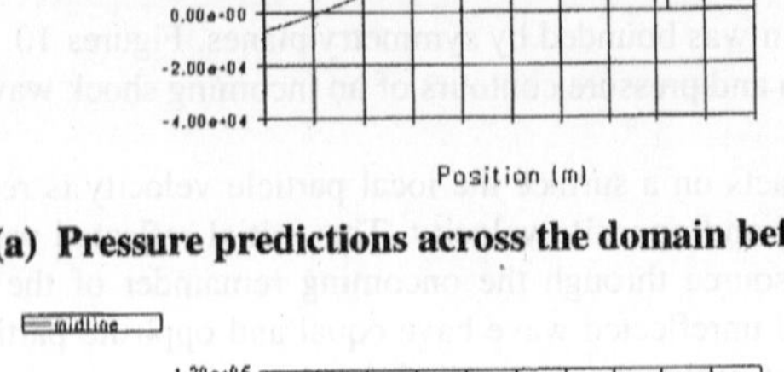

Fig. 10.3(b) Pressure predictions across the domain after reflection

A later test case demonstrates how the magnitude of the wave decays with distance, causing the size of the shock wave as it strikes the wall to be considerably less than when it was first introduced into the domain. Shock magnitudes of 1 bar (Case A) and 10 bar (Case B) were used as test cases, but decayed to wall impingement overpressures of about 0.69 bar and 8.93 bar respectively. Figure 10.3(a) shows the profile of the shock as it approaches the wall. The step rise in pressure across the shock front can be seen, along with the pressure decay in the wake of the wave extending back towards the left side of the domain. Figure 10.3(b) shows the reflected shock wave propagating back towards its source. It should be noted that the overpressure peak occurs in the immediate vicinity of the wall and, thus, the reflected shock has a lower pressure than that experienced at the wall. Additional models with different angles of incidence have been produced. In all models the overpressure value at the wall on wave impact along with the wave reflection angle have been correlated to empirical results from standard texts, and are shown below.

Table 10.1 Blast reflection validation

	Standard Text		Predicted	
Initial pressure at source	Over Pressure (Pa)	Reflection Angle (°)	Over Pressure (Pa)	Reflection Angle (°)
Case A normal incidence	1.72×10^5	–	1.73×10^5	–
Case A 20° incidence	–	20.0	–	20.0
Case B normal incidence	9.93×10^5	–	9.93×10^5	–
Case B 20° incidence	–	10.7	–	9.5

The accuracy of the reflection angle predictions were found to be relatively mesh independent, whereas the overpressure predictions were not. Mesh adaption criteria, based on pressure gradient, have been developed to mitigate this mesh sensitivity so as to optimize the results. The Case A blast predictions all performed well against available data, whereas simulating blasts at higher pressures typically, although not at 20 degrees, proved to be less accurate but satisfactory. Triangular meshes were investigated and found to produce less accurate results, although this type of mesh will probably be required for more complex domains and as such investigations are currently being undertaken to develop a triangular mesh methodology.

10.2.2.2 Wave propagation

The next stage of the validation was to examine the decay of a shock wave propagating in free space. For this a two-dimensional axisymmetric model was constructed so as to represent a spherical domain. A blast was initiated, using a burn model, at the centre of the domain and the developed pressures were recorded. Again, the model was created using a quadrilateral mesh based on an unsteady, compressible, and inviscid solution. A pressure prediction approximately halfway through the simulation is displayed by Fig. 10.4. Note the low pressure region at the centre of the spherical domain. This is due to the outward movement of the air surrounding the explosion.

The predictions correlated well against standard texts, which are shown below.

Table 10.2 Blast propagation validation

Radial distance m	Peak Pressure bar		Arrival Time ms		Duration ms		Impulse bar.ms	
	Predicted	Empirical	Predicted	Empirical	Predicted	Empirical	Predicted	Empirical
0.3727	10.924	10.670	0.180	0.186	0.185	0.194	0.389	0.509
0.7438	2.283	2.020	0.675	0.680	0.436	0.440	0.325	0.291
1.7617	0.347	0.315	2.983	2.920	0.900	1.046	0.132	0.119

Figures 10.5 and 10.6 display the transient pressure predictions at radial distances of 0.37 m and 0.74 m respectively. It may be observed that in both instances, the predicted peak pressure is slightly lower than that given by standard texts. The predictions at a the 0.74 m radius, as shown by Fig. 10.6, can be seen to be more accurate than those displayed at 0.37 m, as shown by Fig. 10.5.

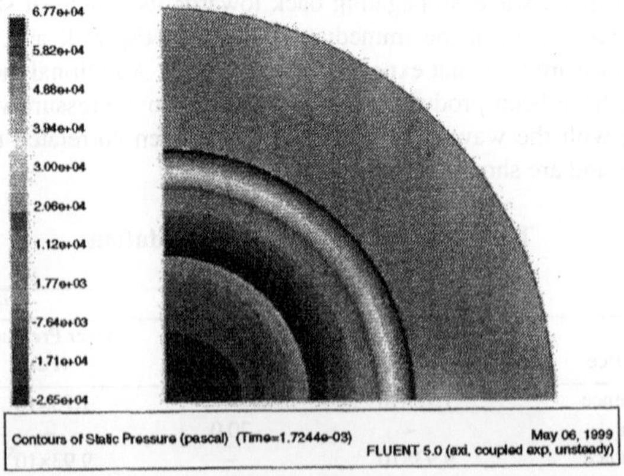

Fig. 10.4 Three-dimensional blast pressure predictions

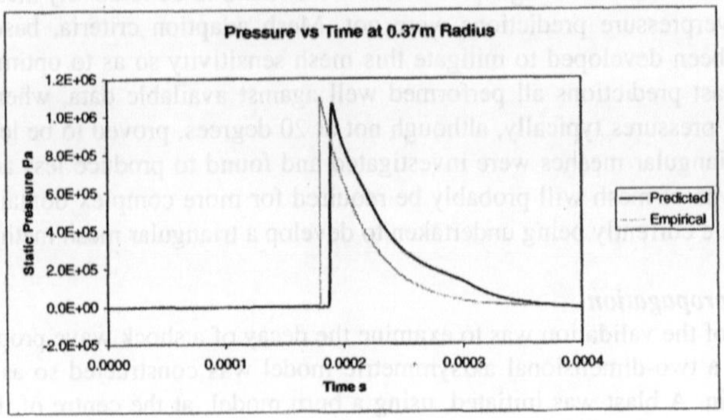

Fig. 10.5 Transient pressure at 0.37 m radius

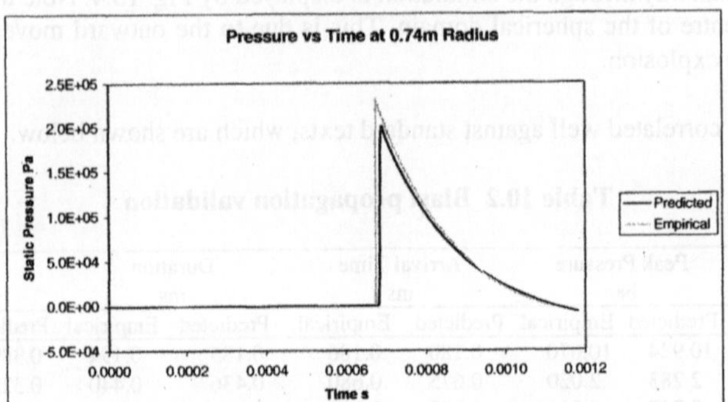

Fig. 10.6 Transient pressure at 0.74 m radius

According to standard texts the radius at which the shock wave first forms is, in this instance, 0.37 m. What is clear from the results shown by Fig. 10.5 is that the shock wave has yet to fully form, as the shape of the pressure profile is not fully representative of an ideal wave, even when ignoring the shift on the time axis. Contrast this with the predictions displayed by Fig. 10.6 where the pressure profile matches closely to the desired shape.

This particular simulation, therefore, does not predict the exact position, and therefore time, of the formation of the shock wave. However, the results are sufficiently accurate considering no mesh refinement was attempted, especially for subsequent points after the shock wave formation. The large error in impulse at the 0.37 m point, shown by Table 10.2, may be attributed to the delayed prediction of the shock wave and therefore the additional effect of the blast wind being incorporated in the impulse. At larger radii this blast wind is not present and hence, the pressure profile and impulse correlate well with known results.

10.2.3 Application example

Having validated an approach to blast modelling more complex structures may be examined. An assessment of blast resistance of a vehicle underbody (1) has been undertaken as a demonstration. The aim was to quantify the kinematic response of the vehicle to a blast. Two vehicle shapes were studied, one with a flat underbody and one with an angled underbody. Identical blasts were released beneath either vehicle and the forces experience by each shape were monitored. Figures 10.7(a) to (d) and 10.8(a) to (d) display the pressure contours developed around the flat and angled vehicle underbody shapes respectively.

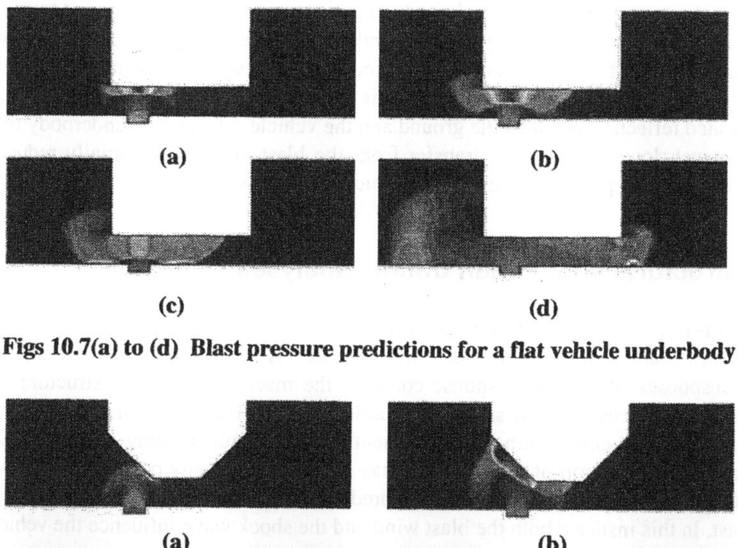

Figs 10.7(a) to (d) Blast pressure predictions for a flat vehicle underbody

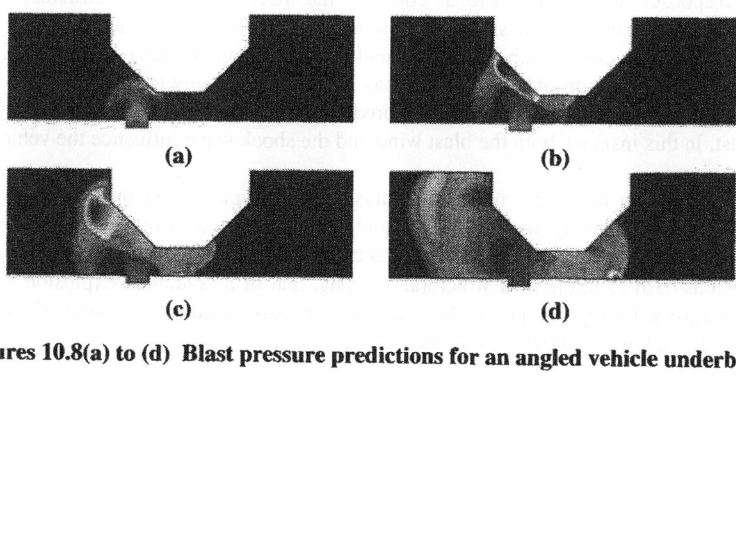

Figures 10.8(a) to (d) Blast pressure predictions for an angled vehicle underbody

The corresponding force predictions are displayed by Fig. 10.9.

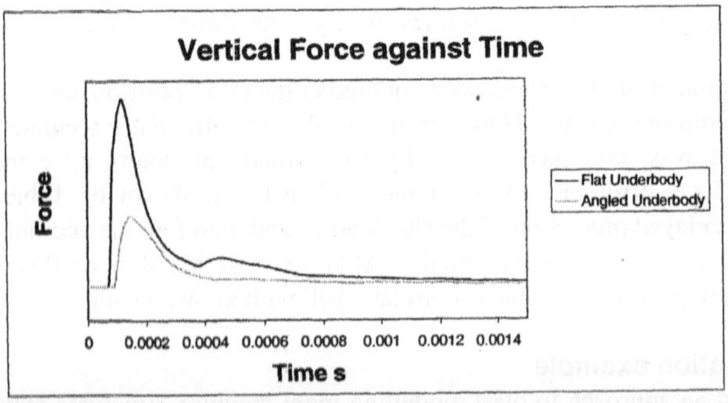

Fig. 10.9 Vertical force predictions for vehicle underbody shapes

It is interesting to observe that Fig. 10.8(a) displays both the shock wave, at the circular extremity of the high-pressure region, and the blast wind continuing to emanate from the source.

Figure 10.9 shows that the angled underbody experiences a lower peak force, approximately 37 per cent of that experienced by the flat underbody. The angled underbody reduces the reflected component of the blast and therefore reduces the developed underbody pressures. In addition, the duration of the angled body force is less than that of the flat underbody, as there is less repeated reflection between the ground and the vehicle. The angled underbody therefore receives a much lower momentum transfer from the blast, thereby potentially reducing the structural damage and potential injury to the vehicle occupants.

10.3 Structural and human trauma analyses

10.3.1 Kinematic and structural response

Once the blast loads have been predicted the next step is the analysis of the kinematic and structural responses. Kinematic response concerns the macro motion of a structure and has two main consequences. First, if significant motion of the structure occurs there may be the secondary effects associated with the subsequent impact on surrounding structures. Second, the acceleration and motion of the structure may cause injury to any occupants or damage to the structure itself. Figure 10.10 shows the predicted kinematic response of a vehicle to a nearby blast. In this instance both the blast wind and the shock wave influence the vehicle.

The output from this kinematic response analysis would typically be the acceleration and motion of the vehicle, along with the associated secondary impacts of the vehicle with the ground. Such data would be used to assess occupant injury or damage to equipment and vehicle systems. An example of a structural analysis, that of a contained explosion within a building, is shown by Fig. 10.11. In this example the blast wind, combustion of explosive products, and shock wave affect the building's response.

The modelling of structural response to blast is more established than the previous aspects discussed. Finite element analysis (FEA) to blast response and crashworthiness problems may be undertaken using a variety of FEA codes. FNC uses a non-linear dynamic code which can be configured to model structural fragmentation as well as deformation.

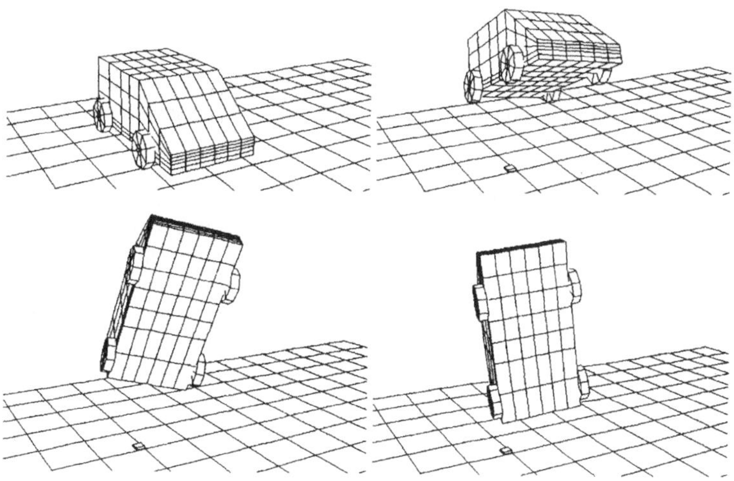

Fig. 10.10 Subsequent kinematics from a near vehicle blast

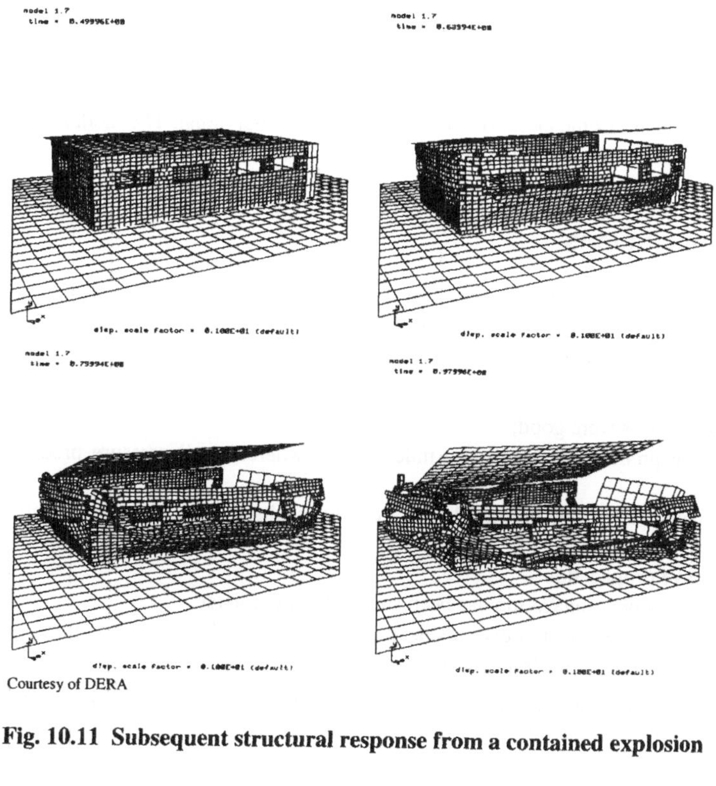

Courtesy of DERA

Fig. 10.11 Subsequent structural response from a contained explosion

10.3.2 Human injury

The final aspect of blast analysis is the determination of the level of human injury. Injury can be caused by primary, secondary, or tertiary blast effects. Primary injury is caused by the direct effects of the high shock pressures associated with blast on the structure. Typical primary injuries are blast lung and eardrum rupture. Secondary injuries are those caused by fragments from the structure, or more typically a weapon system. The final mechanism, tertiary injury, is that caused by the kinematic response of the structure and human. All three of these mechanisms have the potential to cause severe injury, and should be considered in the blast assessment.

Primary injury may be assessed by comparing the predicted blast pressures, obtained from CFD or hand calculations, to standard injury data. The development of injury criteria for blast, particularly with respect to enclosed spaces which cause a complex blast patterns, is particularly important and FNC has been working in this area for a number of years.

Secondary injury requires separate analysis methods to determine the trajectory of structural fragments and their resulting impact. From this it is possible to determine the likelihood of a fragment hit on personnel. The subsequent lethality can then be determined using empirical based injury criteria.

Tertiary injury can be examined at two levels of complexity. First, the kinematic response data can be reviewed and the likely injury levels deduced based on comparison with known injury data. However, if a greater level of confidence in the predictions is required then it is necessary to conduct a more comprehensive assessment. This more complex analysis requires a human modelling technique to examine the motion of the individuals and their interaction with local structures. FNC has developed a technique, which incorporates soft tissue modelling, for simulating humans in such environments to provide an assessment of injury. Figure 10.12 illustrates the effect of a nearby blast on an individual. The model would receive an input from that predicted by the CFD blast modelling.

10.4 Conclusions

A methodology for shock-wave modelling has been validated by Frazer-Nash Consultancy. The validation exercise examined both wall reflection and free space propagation of a shock wave, and the following conclusions may be drawn:

- Wall reflection
  - Predictions were good;
  - Angle predictions were mesh-independent, whereas wall pressure predictions were not;
  - Triangular meshes produced less-accurate results than quadrilateral meshes in wall reflection.
- Shock wave propagation
  - Shock wave propagation has been successfully predicted;
  - The pressure and impulse levels delivered by the blast were predicted accurately, especially those subsequent to the shock-wave formation.

Future validation work will examine the use of triangular meshes and adaption criteria for use with more complex geometries.

In the wider context of blast mitigation a comprehensive analysis is now possible. Simulation of the actual blast provides accurate data for subsequent analyses of structural kinematic response or deformation, together with any human injury aspects. These techniques may be applied to a variety of scenarios, for example:

- Inadvertent industrial blast prediction for structural safety cases;
- Design of blast mitigating structures, for example buildings and vehicles;
- Human injury prediction and mitigation.

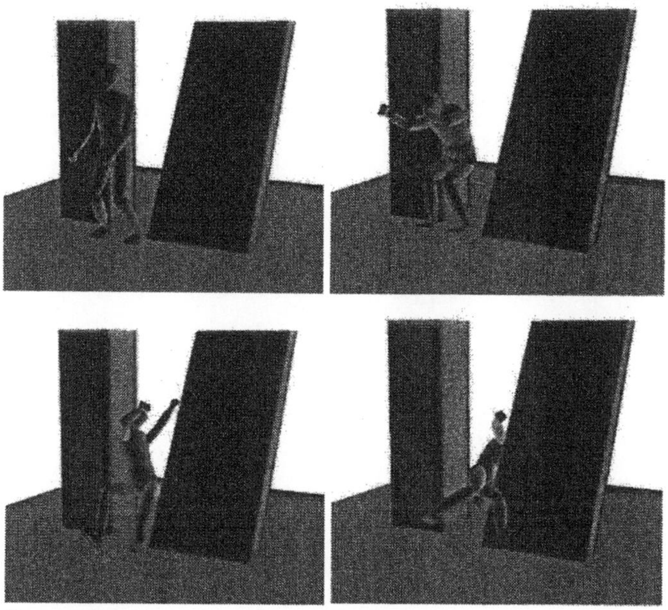

Courtesy of DERA

Fig. 10.12 Model of a nearby blast on an individual

10.5 Acknowledgements

The studies relating to the demolition of the building and the secondary injury to a human were carried out with the support of the Defence Evaluation and Research Agency.

10.6 References

(1) **Dorn, M. R.** (1999) Improving vehicle resistance to blast, *VIIIth European Attack and Survivability of AFVs Symposium*, UK, March.

11

Built Environment Simulations Using CFD

D Woolf and **G Davies**

Synopsis

This chapter is intended to be an overview of a number of technical and design issues encountered within built-environment simulations. Due to the diurnal and seasonal variation of external conditions, a dynamic thermal model developed within Ove Arup and Partners is primarily used to provide the surface boundary conditions for the CFD. The interaction between the numerous supply and extract conditions, the surfaces and internal heat loads (such as occupant and small power loads) often generate complex flow fields with air temperatures close to the ambient condition. It is important to optimize the performance of the ventilation systems, whether they are mechanical or natural. Issues, other than satisfying occupant comfort criteria, may relate to health and safety (e.g. smoke generation and movement from a fire) or other criteria such as moisture gradient in an art gallery and acceptable temperature range for vegetation. The complex flow fields need to be visualized in an easy, interactive way. The biggest challenge, however, is to ensure that the results are both valid and can be explained to a varied audience.

11.1 Introduction

Engineering in the 21st century will encounter very different problems from those seen today. In particular, it is apparent that concerns of sustainable development and pollution will become increasingly central to our work. By finding scientific ways of addressing these issues we can make a significant contribution to mankind. To this end, an understanding of our man-made and natural environment, particularly in terms of the interaction between the two, is crucial. One field that impinges on this understanding, from pollution movement in a city centre to the flow of avalanches, is *fluid dynamics*.

Computer-aided analysis techniques have revolutionalized engineering design in many fields. Computational Fluid Dynamics (CFD) is one such technique which enables the analysis of fluid flow phenomena including heat transfer, mass transfer and chemical reaction. This is achieved by solving the mathematical equations which underlie the physical processes using a mesh based iterative solution method. CFD is principally used to study the behaviour of air and other gases in buildings. In parallel to this, experimental and theoretical methods can be applied. An important example of this is the application of wind tunnel testing to assess wind loads on structures.

The benefits include:

- Rapid and economical evaluation of design performance;
- Parametric and comparative studies to improve designs;
- Identification of important but hidden flow features;
- Simulation of experiments too hazardous to set up; and
- Cost-effective alternative to full or reduced scale testing.

In the building industry CFD is used to assess thermal comfort, ventilation effectiveness, occupant health, energy use, and also for specialist analyses (e.g. external flows, smoke movement, gas dispersion). There are a large number of other principal design fields in which CFD has proved a useful tool. These include automotive (e.g. aerodynamics, HVAC, underbonnet flows), aerospace (e.g. gas turbines, cabin ventilation), industrial applications (e.g. manufacturing, pharmaceutical, process), water and civil (reservoirs, spillways, buoyancy driven flows), mechanical (e.g. pumps and heat exchangers), and the oil and nuclear industries.

In order to examine some of the CFD issues relating to the building industry, the two following sections detail different projects. The first section investigates the various balances within an igloo environment and the second assesses the detailed air flow patterns and temperature distributions in a large-scale 'greenhouse' project in Cornwall.

11.2 The igloo

In Arctic regions, igloos are constructed out of wind-packed dry snow with access gained through a 'door' into a trench area (1). Figure 11.1 shows the CFD geometry representing a typical igloo sized for two persons. At night the igloo is sealed except for a small ventilation hole to allow fumes to escape and fresh ventilation air to enter. An oil lamp continuously burns, providing heat input into the space and to act as a ventilation signal.

In a cold climate, the level of fabric insulation and the volume of infiltration and ventilation air are major factors in determining internal conditions. The design aim is, therefore, to maximize the thermal resistance of the fabric and to minimize air exchange with the outside. In an igloo, these exchanges become critical to the survival of the occupants.

This study uses CFD to determine the heat, moisture, and concentration distributions as well as the air movement within a two-person igloo with an outside air temperature of -30 °C. The igloo is 2.7 m in diameter and has one 400 W burner.

There are many characteristics of the igloo environment applicable to *all* built environments. For instance, increased velocities and CO_2 levels are evident in a warm plume above the burner. This can also be observed for an occupancy-generated warm air plume. The burner also acts as a large O_2 sink, again analogous to an occupancy effect. CFD was used to investigate these balances in spatial terms.

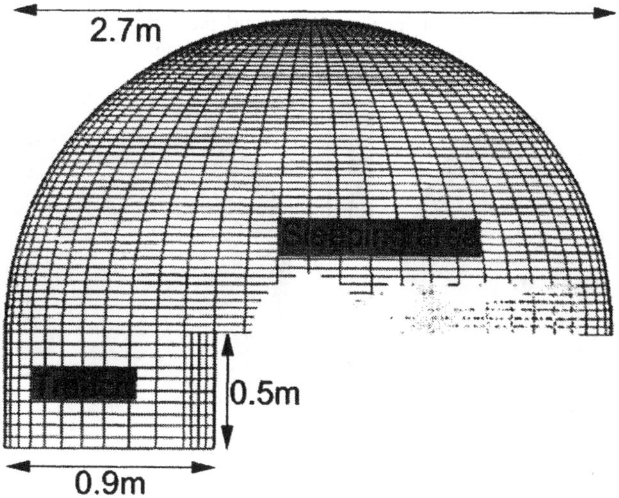

Fig. 11.1 CFD geometry

The extremely cold ventilation air drops into the trench and dilutes CO_2 levels (see Figs 11.2 and 11.3). The sleeping area, on a shelf above the trench, is warmer but also dryer in most locations (relative humidity is a function of the temperature and moisture content of the air). The moisture content is affected by, for example, cooking and respiration.

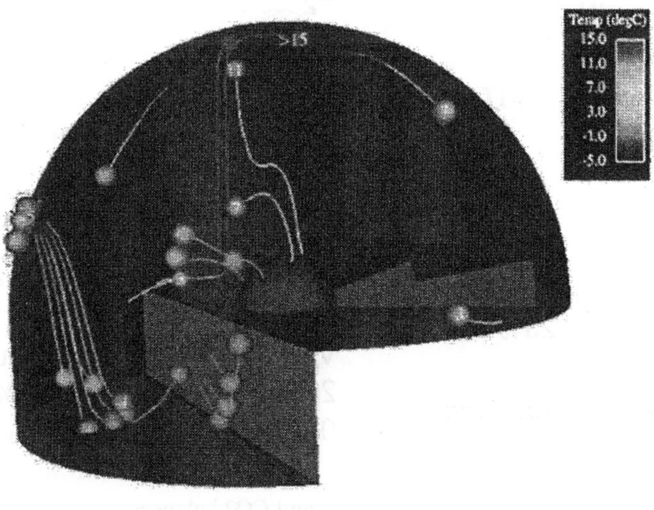

Fig. 11.2 Particle traces showing air movement with temperatures

The benefits of having a trench are evident as it not only provides access through the door, which is eventually 'sealed', but also bounds a cold sink of fresh air, the occupants sleeping on a warm shelf above. The oil lamp is an important element in the thermal balance but it can also act as a detector to see whether there is sufficient oxygen to breathe. Figure 11.4 shows the energy, O_2 and CO_2 balances in an igloo environment. It should be noted that although indigenous designers make use of local materials and methods, valuable lessons should be learned from these designs. The shape of an igloo is no accident.

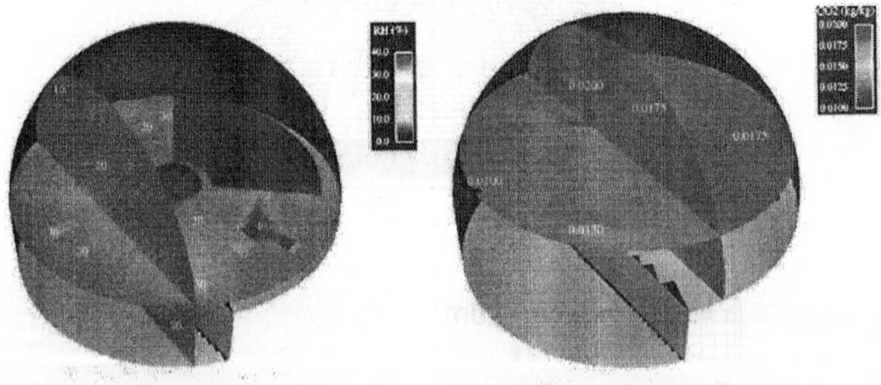

Fig. 11.3 Air relative humidity and CO_2 concentration (horizontal and vertical sections)

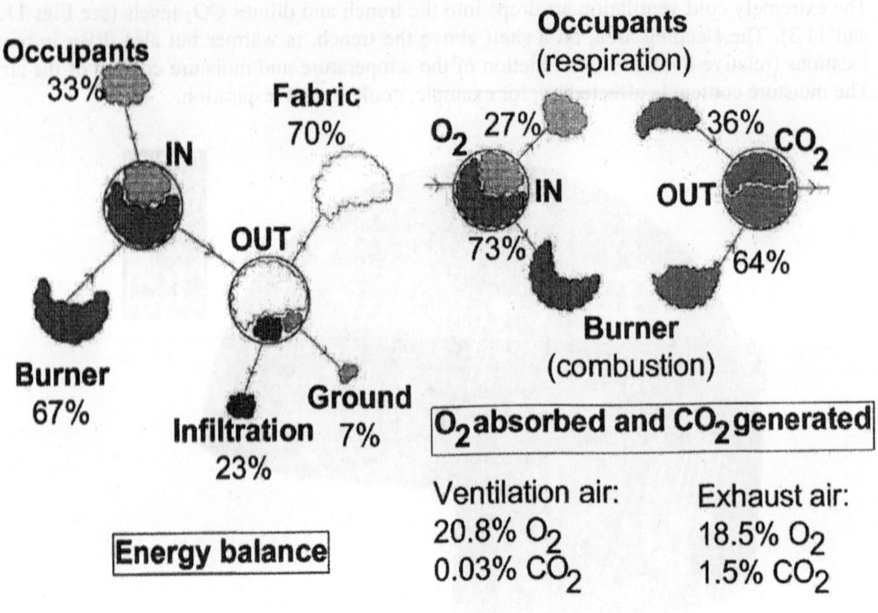

Fig. 11.4 Energy, O2, and CO2 balances

11.3 The Eden Project, Cornwall

The Eden Project consists of a pair of large 'biomes', a collection of dome-shaped greenhouses, which will house plants from all over the world in artificially controlled climates. The biomes are located close to St Austell in Cornwall and are almost one kilometre long and 60 metres high. They are split into two zones representative of a rainforest or the 'Tropics' and Mediterranean or 'Temperate' climatic regions.

The Tropics Biome (see Fig. 11.5) is made up of four partial domes enveloping a volume of approximately 325 000 m^3. These domes are set against a quarry face with the slope increasing in angle to the rear edge of the space on the north side of the Biome. Internally, the quarry walls are south-facing and so high solar gains are possible in the summer. This 'envelope' was split up into a large number of connecting 'finite volumes' or cells, forming a computational mesh which is representative of the internal geometry within which the fluid dynamics equations are solved. The model contained over 440 000 cells. The mesh strategy is to apply the greatest number of cells in the regions where there is the largest velocity and temperature gradients, such as close to the surfaces and discrete supply air nozzles. This strategy enables the airflow to be simulated more accurately.

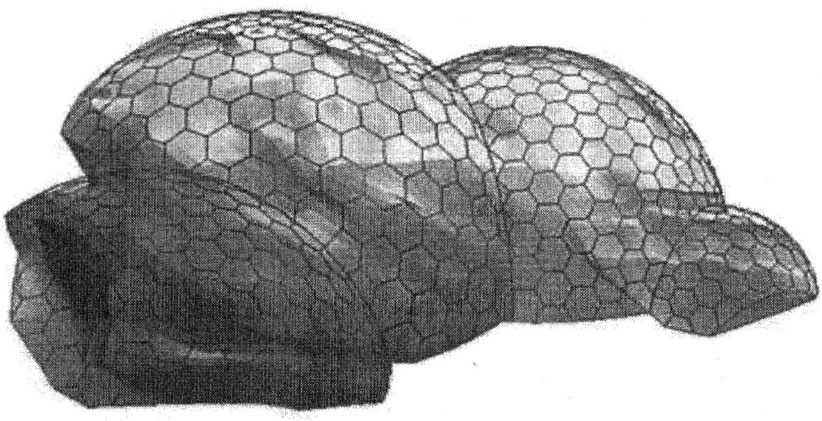

Fig. 11.5 Tropics Biome topography

The purpose of the study is to understand the 'performance' of the space having taken a 'snapshot' in time to simulate design internal and external conditions. The performance is dependent on the effectiveness of the natural ventilation openings in summer and mechanical systems in winter in achieving the design conditions. The main design criteria relate to the maximum allowable air temperature in the vegetation or 'planting' zone and minimum temperature in winter.

It is generally accepted in modelling the built environment, that the differences in the time scale between convective airflow movement and surface temperature changes (typically in the order of seconds to hours respectively), allows the superposition of steady surface temperatures in a CFD model. These surface temperatures are obtained from a dynamic thermal model that may also provide the 'time' of analysis depending upon the objectives of the study. The thermal model, used in parallel to the CFD, assesses the macroscopic performance of the biomes over a twelve-month period on an hour-by-hour basis.

The available input data for the model was in the form of two-dimensional contour levels for the site topography and three-dimensional points and lines (hexagons) describing the roof surface. The contour levels were used to create a series of 'splines' having different Z-components. These were used to create two-dimensional shells describing the ground. The hexagons were split into smaller two-dimensional 4-sided shells to complete the surface definition. Supplemental SAMM (Semi-Automatic Meshing Methodology) software was then applied to create the finite volume mesh. The mesh consisted of a layer of prism cells from the surface to a nearby sub-surface and then a Cartesian mesh within the volume bounded by the sub-surface. The sub-surface 'cut' the Cartesian mesh creating a number of polyhedral cells (from hexahedral) at the sub-surface interface. Some problems were experienced with 'unresolved' cells at this interface but recent improvements to this software should negate these problems in the future.

One of the main challenges is to ensure that the results are both valid and can be explained to a varied audience. Key design issues include items such as optimal grille and sensor locations, moisture content gradient, as well as the more usual air temperature and velocity distributions. In the biomes, very complex three-dimensional flows were generated in both the summer and winter analyses. It is possible to show data, such as velocity vectors, on a two-dimensional slice but this will often not tell the full story and could, in some instances, even be misleading (see Fig. 11.6).

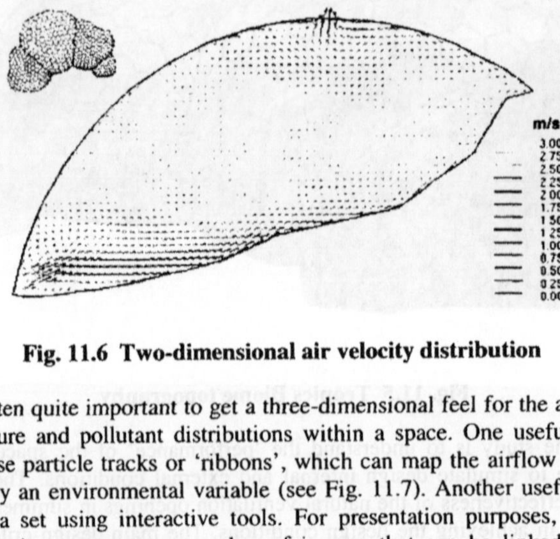

Fig. 11.6 Two-dimensional air velocity distribution

It is therefore, often quite important to get a three-dimensional feel for the airflow movement and air temperature and pollutant distributions within a space. One useful post-processing technique is to use particle tracks or 'ribbons', which can map the airflow movement whilst being coloured by an environmental variable (see Fig. 11.7). Another useful technique is to interrogate a data set using interactive tools. For presentation purposes, one of the most valued techniques is to generate a number of images that can be linked together into an animated sequence (see Fig. 11.8).

The results from this study provided confidence in the decisions that were being made in the development of the design. For instance, site 'buildability' issues could be addressed with adjustments made to the number and location of supply nozzles for the warm air system in winter. This also led to large savings in the amount of mechanical plant required.

11.4 Reference

Cook, J. (1996) Architecture indigenous to extreme climates, *Energy Buildings*, **23**, 277–291.

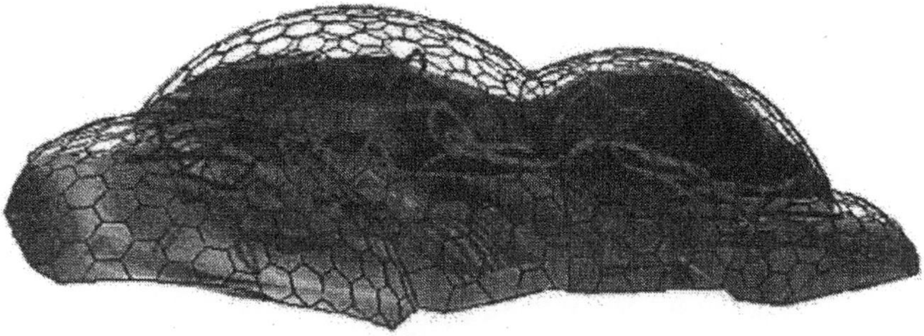

Fig. 11.7 Particle tracks for the summer analysis

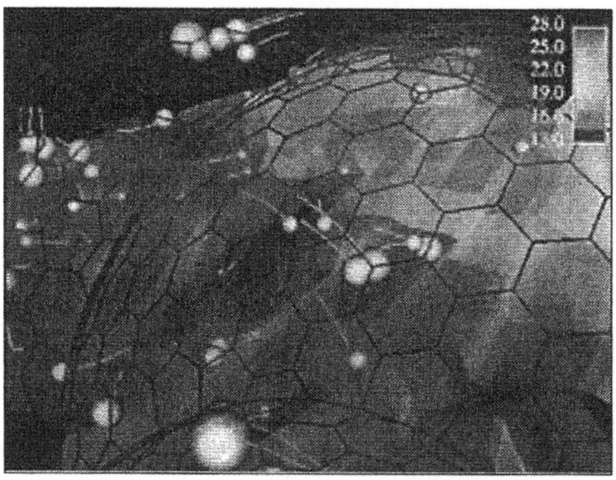

Fig. 11.8 Part of an animated sequence with moving particles

12

Using CFD in the Design of Electric Motors and Generators

S J Pickering, D Lampard, J Mugglestone, M Shanel, and **D Birse**

Synopsis

Convection heat transfer and ventilation in rotating electrical machines is poorly understood, yet thermal constraints impose limitations on the power capability of many motors and generators. CFD has the potential to provide understanding of the cooling process and aid in the design of improved machines.

This chapter describes research done by The University of Nottingham in collaboration with ALSTOM Electrical Machines Limited to investigate airflow and heat transfer in an electric motor. The particular area of interest was the cooling of end windings in a large induction motor. CFD modelling was developed, using the commercial CFD code Fluent, and compared with results from an extensive programme of experimental measurements.

The chapter will describe the problems that were faced in applying CFD to electrical machines in modelling:

- The complex geometries of the end winding coils;
- The interaction between the rotor and stator.

As a result of the work, confidence has been gained that has led to the more general use of CFD in machine development at ALSTOM Electrical Machines Limited. The benefits that CFD has brought will be described.

12.1 Introduction

One of the major limitations in the drive to improve the performance of rotating electrical machines is the temperature limitation of the materials used in the construction. In the drive to manufacture more competitive machines, effective design requires accurate computation of local temperatures. This in turn, requires accurate modelling of the thermal performance of the machine which requires detailed knowledge of the thermal boundary conditions, including the convective heat transfer coefficients. Despite the importance of convective cooling in electric motors and generators, detailed data on heat transfer coefficients are not generally available and there are few published references (**1**)–(**3**).

In a recently completed research project at The University of Nottingham, detailed experimental measurements of heat transfer on the end windings and internal frame surfaces of a totally enclosed fan-cooled (TEFC) induction motor were obtained. As part of this study the end region of the motor was also modelled computationally to ascertain the usefulness of CFD as a possible design tool. This chapter describes the results obtained from the CFD modelling and compares them with the experimental measurements. Following on from the success of the modelling, CFD has since been implemented within an industrial design office environment at ALSTOM Electrical Machines Limited and this will be described.

12.2. Motor configuration tested

The test motor was two-pole, high-voltage, TEFC induction motor of approximately 200 kW rating with strip wound coils. A diagram of the end region is shown in Fig. 12.1. Figure 12.2, taken from the CFD model, shows a more detailed picture of the end region configuration. A test rig was constructed, described below, to simulate the end region of the motor in which air circulates to transfer heat from the rotor and stator to the frame.

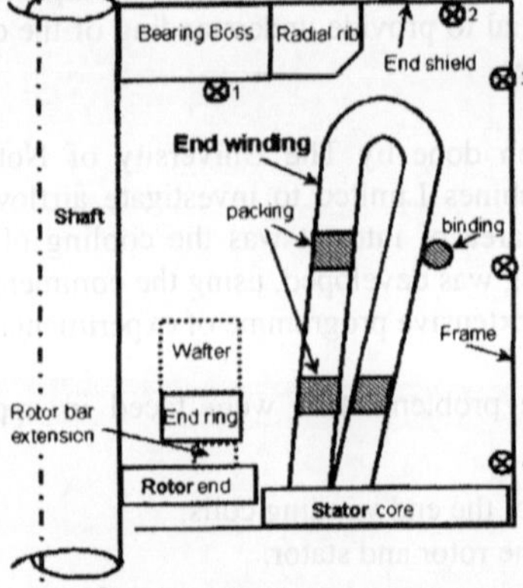

Fig. 12.1 Diagram of cross-section of end region of test motor

12.3 Capabilities of CFD

General-purpose CFD is now available to model fluid flow by solving, numerically, the fundamental fluid flow equations (Navier–Stokes equations). However, there are two particular features of the air flow in the end region of an electrical machine that make it difficult to model.

1. The motion of the wafters and rotor bar extensions generate an inherently unsteady air flow due to interaction with the stationery end winding. This interaction between the rotor and the stator, creates a complex flow pattern where the mean flow exhibits

periodic fluctuations as the wafters and rotor bar extensions sweep past the end winding coils. This unsteady air flow has important influences on convective heat transfer.

2. The end winding has a complex open geometry with many small gaps in between the coils. This detail must be modelled accurately in order to provide detailed and accurate information on local heat transfer coefficients around the endwinding.

Fig. 12.2 End winding and rotor configuration (detail from CFD model)

12.3.1 Modelling rotor-stator interaction

The best method of modelling rotor stator interaction is to use the *sliding mesh* technique. In this method, the rotor and stator are modelled using separate meshes. These slide relative to one another and a time dependent solution is obtained to represent the interaction in full.

However, at the time that this study was undertaken the CFD software used (FluentUNS) did not have the capability to use the sliding mesh technique and the *rotating reference frames* method was used instead. In this method the rotor and stator are still modelled using two meshes. But it is assumed that there is steady flow at the interface between the rotor and stator and a time-averaged solution is adopted. This is much more economical in computing time, but does not give a true representation of the unsteady nature of the rotor-stator interactions.

12.3.2 Modelling the geometry

It is only during the past few years or so that the widely used commercially available CFD codes have had the capability to model complex geometries with unstructured terahedral cells. Previously, regular hexahedral block type meshes had to be used with the constraint that rows of cells had to be continuous through the computational domain. This meant that complex geometries such as end windings could not practically be modelled. The use of FluentUNS, an unstructured CFD code, greatly facilitated the modelling of the end winding geometry as shown in Fig. 12.2.

12.4 CFD modelling of test motor

The motor test rig had eight wafter blades and, as a result of symmetry, only a 45 degree sector of the end region had to be modelled. This was done using the commercial CFD code FluentUNS Version 4. The 45 degree sector was modelled using about 500 000 cells and represented all of the main geometric features of the test motor. The rotor stator interaction was modelled using the multiple reference frames approach which assumes a steady flow at the interface between the rotor and stator.

Several turbulence models were investigated including the k–ε model and the k–ε RNG model. It was found however, that the popular k–ε model performed the best, despite its acknowledged weakness in modelling highly swirling flows.

12.5 Experimental measurements

A model of the end region of the test motor was constructed as shown in Fig. 12.1. The end windings from a real machine were used and mounted within a perspex cylinder to simulate the motor frame. The end of the rotor, end ring, wafters and the shaft were simulated and driven directly by a small motor mounted beneath the rig. For most of the tests the rotor was driven at 1700 r/min.

Air flow measurements were made within the test rig using hot wire anemometry to give measurements of velocity magnitude. This enabled air flow speed and direction to be measured. One end winding coil was instrumented with 25 miniature thin film heat flux gauges and thermocouples. A current was passed through the coil to provide a thermal input. Local heat transfer coefficients h_{local} were then deduced from the local measurements of heat flux $\dot{q}''$ and surface temperature, $T_{surface}$, and bulk air temperature, T_{bulk}, using the formula:

$$h_{local} = \frac{\dot{q}''}{T_{surface} - T_{bulk}}$$

An overall heat transfer coefficient was also obtained for the whole end winding coil based on the heat input, deduced from electrical power input, and average surface temperature, based on measurement from thermocouples located at points on the coil surface. Figure 12.3 shows the locations of the heat flux gauges on the end winding coil. Figure 12.1 shows the locations of the heat flux gauges on the inside of the end shield and frame of the test rig that were used to calculate the heat transfer coefficients on the inside of the motor frame.

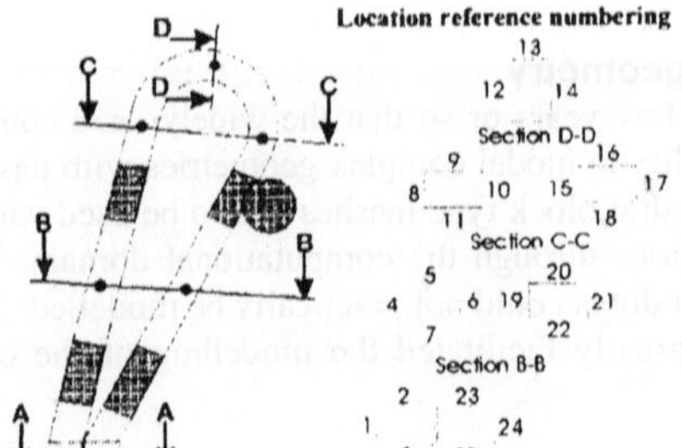

Fig. 12.3 Location of heat flux gauges on end winding

12.6 Results

12.6.1 Air flow

Figure 12.4 shows a detailed comparison of the CFD predictions and measurements of air velocity magnitude on an axial traverse taken between the wafters and the end winding. This clearly shows that the overall air flow is well predicted over a range of rotational speed and that the velocity magnitudes in this strongly swirling flow are generally predicted to within 10 per cent of the measurements.

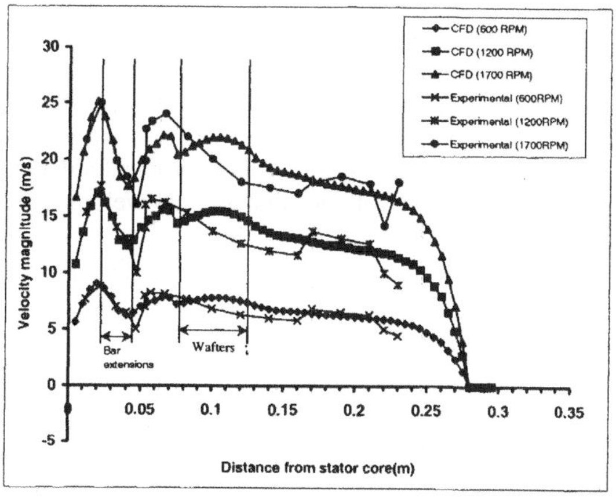

Fig. 12.4 Air velocity within the end region – (comparison of CFD and measurement)

In this study the main aim was to obtain a detailed understanding of heat transfer and so only limited air flow measurements were made. However, in an earlier study of the effects of end winding porosity on air flow, more detailed air flow measurements were made in the three component directions (5). Again, it was found that the CFD predictions agreed well with measurements in all three component directions.

12.6.2 Overall heat transfer coefficients

Figure 12.5 shows a comparison of CFD predictions and measurements of overall heat transfer coefficients on the end winding coils and inside the motor frame in the end region. In each case the CFD predictions are lower than the measurements and are within 20 per cent. The comparison is made over a speed range from 600 to 1700 r/min. The variation of heat transfer coefficient with speed is very well predicted, and the predictions of the exponent for the variation of heat transfer coefficient with speed are almost identical to the measured values.

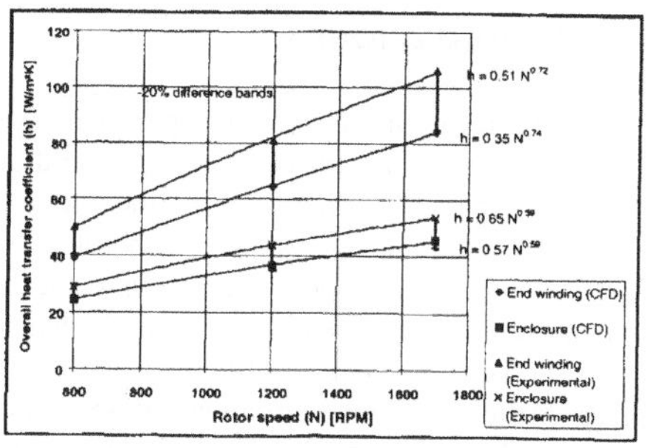

Fig. 12.5 Overall heat transfer coefficients on the end winding and inside frame (comparison of CFD and measurement)

12.6.3 End winding heat transfer coefficients

Figure 12.6 shows the variation of local heat transfer coefficient around an end winding coil. The location is on the inner face of the end winding (positions 4, 5, 6, and 7 in Fig. 12.3) opposite the wafter blades. In this position there is a strongly swirling flow on the inside surface of the end winding. Some air passes down the gaps between the coils and separates from the downstream side of the coil creating a region of recirculation. Consequently, it is expected that the heat transfer coefficients would be higher on the side of the coil facing the swirl direction of the air flow than on the downstream face. This can be seen from Fig. 12.6. The region on the rear of the coil (position 6) has a low air flow and this has the lowest heat transfer coefficient The CFD predictions of heat transfer coefficient are all within 30 per cent of the measured values and the variations with position are very well matched.

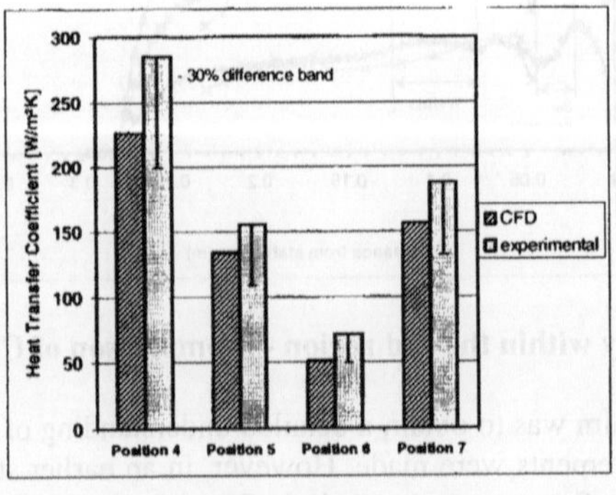

Fig. 12.6 Variation of heat transfer coefficient around an end winding coil (comparison of CFD and experiment)

It was also found that the CFD predictions of heat transfer coefficient variations along a coil were within 30 per cent of the measured values at any location. Fuller details of these results and also comparisons between CFD and experiment of the effect of various geometric variations are given in other publications (6)–(8). Heat transfer coefficients on the frame of the motor were also modelled and found to be within 30 per cent of experiment (8). Comparisons of predicted windage loss were found to be within 20 per cent of experimental measurement (8).

12.7 Comments on CFD predictions

In general, the agreement of the CFD predictions with experiment is surprisingly good. It would not be expected that measured heat transfer coefficients would agree exactly with the CFD predictions as there are a number of deficiencies in the modelling. Firstly, the end winding coils were assumed to be rectangular in cross-section with sharp corners. In practice they have rounded corners and the surface is somewhat uneven due to the winding of the insulation around the coils. Rounded corners would, in general, give somewhat higher air flow rates and the uneveness in the surface means that it is difficult to position the heat flux gauges in truly representative positions. This, combined with manufacturing variations between coils, means that there would be expected to be variations in measured heat transfer coefficient between heat flux gauges mounted at the same position on different end winding coils.

However, perhaps the most important deficiency in the modelling is as a result of the use of the rotating reference frames method for modelling the rotation of the rotor. This assumes a steady flow at the interface between rotor and stator. In reality, the flow is unsteady and the end winding coils would experience fluctuations in air flow as each wafter sweeps past a coil. This unsteadiness would increase the heat transfer coefficient and so it is only to be expected that the CFD predicted heat transfer coefficients are lower than the measured values.

Another factor is the CFD turbulence model. The k–epsilon model used in this study is known to be inaccurate in strongly swirling flows as it assumes that the turbulence is isotropic. The Reynolds Stress model is more appropriate for swirling flow but was not available in FluentUNS with rotating reference frames at the time this study was undertaken. However, it has the disadvantage of being more computationally intensive.

12.7.1 Potential for design

The results of this study have shown that, even with known inadequacies in the CFD modelling, heat transfer coefficients within the end region of a TEFC induction motor on the end windings, frame, and end shield can be predicted to within 30 per cent accuracy. The predicted heat transfer coefficients are generally lower than those measured and so, for design purposes would give results that would be in error on the safe side; i.e. they would predict temperatures higher than would actually be achieved. CFD models of ventilation systems would thus yield heat transfer coefficients that could be used as boundary conditions to thermal models of complete machines. And these could be expected to give much more accurate results than has hitherto been possible.

A particular advantage of CFD is that much more detailed information can be obtained than would be available from experimental measurement, particularly in complex geometries such as end windings where access for measurement is very restricted. The detailed understanding of the flows and heat transfer that CFD can give should give insights to enable improvements in machine design to be made to enhance thermal performance. CFD techniques are rapidly becoming more cost-effective in the development of new machines than more traditional methods of experimentation.

The CFD modelling carried out in this study was done on workstations that are typically used in industrial design offices today and the computation of one model typically took hours. Sliding mesh techniques are now commonly available and would be expected to give more accurate results. However, as they require solutions that are time dependent they will generally mean a substantial increase in computational effort required.

12.8 Application of CFD in design office

In common with many industries, electrical machine manufacturing is feeling the effects of the environmental lobby. This, coupled with the ever-present need to reduce costs and maintain competitiveness, has placed even greater design emphasis on the efficient use of materials. Optimal management of the airflow within the machine to improve cooling is one of the means by which this can be achieved.

The dominating factor, which limits the rating of an electrical machine, is the temperature rise due to internal losses. The effects of high temperatures upon the winding insulation materials determine the limits and the various international standards organizations have set maximum operating temperatures for each grade of insulation. It is the designer's task to maximize the output/volume ratio for a given machine while keeping the winding temperature within the limit.

The heat loss generated within a machine is the combined effect of electrical and mechanical losses. These losses must be transferred to the cooling medium via a two-stage process:

1. Conduction through copper, insulation, and iron to the cooling surfaces;
2. Convection from these surfaces to the cooling medium, which in most cases is air.

In the early days of machine design, particularly where generators were concerned, many machines were slow speed and large in size. It was often the case that natural convection and the fanning effect of the rotor were sufficient to provide cooling. As machines have developed, higher ratings have been required from physically smaller packages. This concentration of losses in a smaller volume has only been possible by greatly improving the ventilation.

The mechanical power required to circulate the air around the cooling circuit is a significant loss in an air-cooled machine and has a significant bearing upon the efficiency of such machines. The designer must therefore make every effort to minimize the windage loss and to maximize cooling circuit effectiveness. With ever decreasing tolerances required on all design calculations and the high costs of getting it wrong, designers are venturing into areas where traditional predictive capabilities on their own are inadequate.

The standard design approach to air circuit modelling has taken the form of resistance network solutions for pressure drop and flow velocity. These techniques have their foundations in manifold theory developed in the late 1950s. Over the years, this theory has been augmented (4) at ALSTOM Electrical Machines Limited to include expressions for rotational effects and the self-generated head loss. This improved formulation was derived through an extensive process of experimentation and correlation. A major drawback with the network approach is that it is inherently one-dimensional, does not have any form of detailed air flow model (incorporating turbulence), and cannot therefore look at the localized cooling effects.

Modern CFD codes, such as Fluent, offer an alternative, attractive, and efficient tool that can be used to explore optimal solutions for given design requirements. Sliding meshes and multiple frames of reference are just some of the advanced features of these packages which make them highly applicable to the rotating electrical machine product.

The high level of discretization within a CFD solution gives detailed information on the localized air flow patterns. Whereas previously the arrows drawn upon a diagram of the air circuit could only indicate the intended path of the cooling air, it is now possible to address the movement of air with greater confidence.

Using the advanced graphical features of CFD codes, hot spots and the mechanisms that give rise to them can now be visualized. Thermal comparisons are an attractive means of validating CFD results as thermal tests are performed on most production jobs as a matter of course and increased detail can be obtained in the tests for minimal extra costs.

Time is at a premium on production work, with short lead times and high penalties for lateness. Beyond that which is required by the specification, there is very little scope to perform detailed air measurements on the majority of products shipped. Without this baseline of detailed test information, CFD as a design office tool is generally used for relative studies.

Within ALSTOM Electrical Machines Limited, CFD is finding a place right at the heart of the development process. New and untested developments, components with particularly complex geometry (strip on edge field winding) or which are located in areas of extreme turbulence (field winding supports) have an obvious affinity with the CFD approach. Some structures, such as end-windings have proved to be beyond any previous predictive capability and yet are more suited to CFD methods.

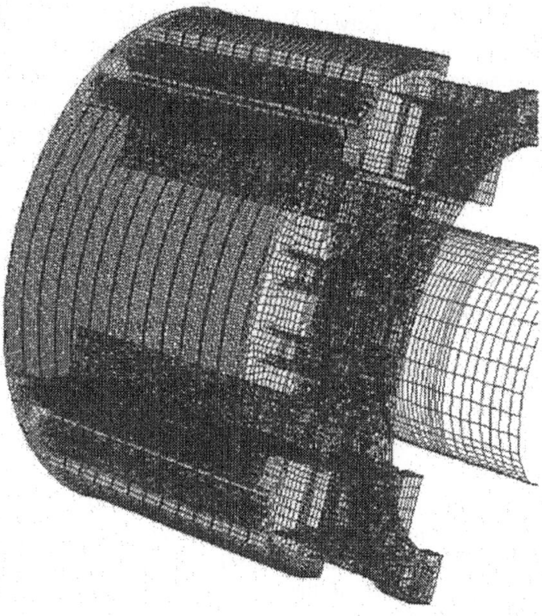

Fig. 12.7 CFD mesh of rotor of large electrical machine

From simple wafters to more complex aerofoil section, fans are components that lend themselves to shape optimization via CFD means. This is an area that has received considerable attention and with significant gains on some of the larger products. Indeed on certain products it has once again been possible to dispense with the use of fans altogether by harnessing the geometry of the machine to assist cooling.

Other areas to which CFD has been applied include:

- Cooler design, looking at the efficiency of the heat exchanger mechanism used to dissipate the heat from the cooling medium;
- Ventilation duct geometry, investigating the effect of change to the through paths within the core components of the machine;
- Envelope reduction, minimizing the space requirement of the channels used to transfer air from the cooling surfaces to the heat exchanger.

As available hardware resource improves, it is now possible to consider a multidisciplinary approach to analysis. Within ALSTOM Electrical Machines Limited, CFD grids are transferred to Finite Element Analysis (FEA) packages on a regular basis, for additional thermal and structural analysis. Thermal constraints are also transferred with negligible user intervention. This provides a seamless airflow/thermal predictive capability.

What does the future hold for CFD at ALSTOM Electrical Machines Limited? The prospect of coupled field analysis, parallel processing, and improvements in hardware suggest that CFD will become an integral part of product development and design.

12.9 Conclusions

The research undertaken at The University of Nottingham has demonstrated that currently available CFD software can be used to model ventilation flows within the end region of an electric motor. The modelling gave air flow predictions to within 20 per cent and heat transfer predictions to within 30 per cent of measurement. And valuable insights into the phenomena taking place were gained. The confidence gained from this research has prompted the use of CFD in the design office at ALSTOM Electrical Machines Limited the prospect is for CFD to become a tool that is integrated into the design process.

12.10 Acknowledgement

The authors gratefully acknowledge the support of EPSRC and ALSTOM Electrical Machines Limited in this research.

12.11 References

(1) **Luke, G. E.** (1923) The cooling of electric machines, *Trans AIEE*, **42**, 636–652.

(2) **Schubert, E.** (1968) Heat transfer coefficients at the end winding and bearing covers of enclosed asynchronous machines, *Elektrie*, **22**, 160–162.

(3) **Oslejsek, O.** (1972) The cooling of the end windings of small enclosed electric machines, *Elektrotech Obzor*, **61**, 548–556.

(4) **Carew, N. J.** (1969) Flow distribution and pressure drop in salient pole electrical machines, *Proc. Instn Mech. Engrs*, **184**, 62–69.

(5) **Mugglestone, J., Lampard, D.,** and **Pickering, S. J.** (1998) Effects of end winding porosity upon the flow field and ventilation losses in the end region of TEFC induction machines, *IEE Proc. Electr. Pwr Appl.*, **145**, 423–428, September.

(6) **Mugglestone, J., Pickering, S. J.,** and **Lampard, D.** (1999) Prediction of heat transfer from the end winding of a TEFC strip-wound induction motor. Proceedings of *IEMDC'99*, IEEE International, *Electric Machines and Drives* Conference, Seattle, USA, pp. 484–486, May.

(7) **Mugglestone, J., Pickering, S. J.,** and **Lampard, D.** (1999) Effect of geometry changes on the flow and heat transfer in the end region of a TEFC induction motor, Proceedings of the Ninth IEE International Conference on *Electrical Machines and Drives*, Canterbury, UK, pp. 40–44, September.

(8) **Pickering, S. J., Lampard, D.,** and **Mugglestone, J.** (2000) The use of computational fluid dynamics in the thermal design of rotating electrical machines, Accepted for publication in *Acta Polytechnica, J. Advanced Engng,* No.4.

13

CFD Modelling of Two-phase Mixing/separation Flow

W M Dempster

13.1 Introduction

The modelling of general two-phase fluid systems and the subsequent predictions using computational fluid dynamic techniques has advanced considerably in the last thirty years primarily due to the research efforts of the nuclear safety community. This work has established a sound basis from which a range of models has been developed depending on the scale and detail of the phenomenon that is required. The most generally applicable is the Two Fluid Model which consists of a set of multi-dimensional volume averaged transport equations for each phase or component of the flow and a set of subgrid models or closure equations to account for interaction at the phase interface.

An in-depth discussion on the modelling approaches to multiphase flows is delivered in the short courses by Hestroni and Yadigaroglu (1), however, an overview is presented by Ishii and Mishima (2). The development of the model has now reached a point where it has found regular application in the nuclear and oil industries in specialized codes and is also found as part of the standard capabilities of the commercial CFD codes. Unfortunately, even though the two fluid model is applied on a regular basis the application to general problems is still problematic, particularly when the liquid fraction lies in the range 0.15 to 0.9. At the extremes of this range lie the spray regime at the lower end and the bubble regime at the higher end of liquid fraction where a number of successful models have been developed. In this problematic range of liquid fractions, in both horizontal and vertical flow conditions the flow can become intermittent with large time scales and complex phase interactions for which only limited understanding exists. This results in a wide range of practical fluid separation and mixing problems that are currently difficult to predict for. It is this class of problem that is addressed in this chapter and attempts to show what is possible with the standard multi-dimensional two-fluid model applied to a combined mixing-separation air water process. Both an experimental and computational exercise was conducted and it will be shown that successful predictions are possible. However, it will also be stated that the successful outcome of a prediction exercise is unlikely without prior knowledge of the flow regimes.

This chapter will describe the strategy to solve the problem, explain the model and closure equations and indicate the predictive accuracy of the model.

13.2 Notation

A	Downcomer flow area	m^2
A_i	Interfacial area	m^2
A_{cl}	Pipe area	m^2
C_o	Correction factor (equation 13.5)	
D	Characteristic DIMENSION or pipe diameter	m
D_h	Hydraulic diameter of downcomer	m
D_d	Droplet diameter	m
F_W	Wall force	N
f_W	Wall friction coefficient	
g	Gravitational acceleration	m/s/s
J^*_g	Non-dimensional air flowrate	
J^*_{LP}	Non-dimensional lower plenum liquid flowrate	
J^*_W	Non-dimensional total liquid flowrate	
L	Length of downcomer	m
M_{ik}	Phase interfacial drag	N/m^3
M_{wk}	Phase wall drag	N/m^3
M_W	Total liquid mass flowrate into downcomer	kg/s
M_{LP}	Total liquid mass flowrate into lower plenum	kg/s
M_{lin}	Liquid mass flowrate entering annulus from one connecting pipe	
Q	Volume Flowrate	m^3/s
S	Downcomer gap	m
v_k	Phase velocity	m/s
u	x-coordinate velocity	m/s
v	y-coordinate velocity	m/s
V_{lin}	Liquid velocity entering annulus from connecting pipework	m/s
W	Downcomer circumference	m
x	Mass quality	
y	Fraction of inlet liquid flowing to lower plenum	

Greek symbols

α_k	Void fraction	
ρ_g	Density of air	kg/m^3
ρ_k	Phase density	kg/m^3
ρ_w	Density of water	kg/m^3
σ	Surface tension	N/m
μ_g	Air absolute viscosity	Ns/m
μ_l	Water absolute viscosity	Ns/m
τ_k	Phase shear stress	N/m^2
τ_k^t	Turbulent phase shear stress	N/m^2
δ	Liquid film thickness	m

Subscripts

i	Phase interface
in	inlet

g gas
a air
l liquid
r relative
w water or wall

13.3 Problem description

The mixing-separation problem that is discussed here has a practical application in the nuclear industry and relates to its occurrence during a loss of coolant accident in pressurized water reactors. It has therefore been studied extensively both experimentally and computationally. However, the prediction of the general problem, involving large scale, three-dimensional thermal hydraulic processes remain largely unresolved. Figure 13.1 shows the geometry of the mixing vessel and indicates four pipe connections. Three of these pipes supply a liquid flow to an annular space at the periphery of the vessel. The fourth pipe connection is an exit pathway for the air and entrained liquid. Air flows into the top and flows down though a core region and reverses in a lower plenum to flow into the annular space to mix with the downward flowing liquid.

The result of the interaction of the upward-flowing air and downward-flowing liquid is a complex two-phase flow regime distribution incorporating, wall based liquid film flows, churn flows, and dispersed flows around the annulus. An indication of the type flow pattern that result in the annulus is shown in Fig. 13.2. The type of flow regimes and the extent of the interaction is largely decided by the magnitude of the airflow which also controls the separation of liquid into a flow to the exit pipe connection and a flow to the lower plenum. If the air flowrate is low then all the liquid will flow to the lower plenum and for the experiments conducted flow out of a bottom exit. If the airflow is high enough then it is possible to transfer all of the liquid entering into the annulus to the exit pipe therefore bypassing the lower plenum. However, this latter process results in significant mixing of the phases in the annulus region. Between these two extremes there are a range of conditions where liquid will separate into a flow to the pipe exit and a flow to the lower plenum. The problem therefore, is to predict the extent of liquid flow division between the pipe exit and the lower plenum exit. This problem had previously been studied experimentally by the author and a large database of experimental information was available for comparison with predictions and as an aid in the development of a model.

Since the two fluid model equations have been produced by a volume averaging process it does not know *a priori* what type of flow regimes are occurring. This then requires the development of a flow regime model to identify the flow regime and the appropriate closure relationships for interfacial transfers, (for this problem only momentum transfer occurs and requires the identification of appropriate wall friction and interface drag laws). For this class of problem the question of the extent of parallel experimental investigation and how it would be implemented was not clear. Identification of the flow regimes for such multi-dimensional flows has received little attention and no mechanistic drag models for the anticipated flow regimes exist. Therefore, how the two fluid model can be applied to this class of problem without prior experimentally derived knowledge was not obvious. It was therefore an objective of this study to identify clearly what aspects of closure models could be obtained from the available literature and what required to be obtained from model testing. Approaching the problem in this manner allows the generality and applicability of the two phase models available in commercial CFD codes to be determined.

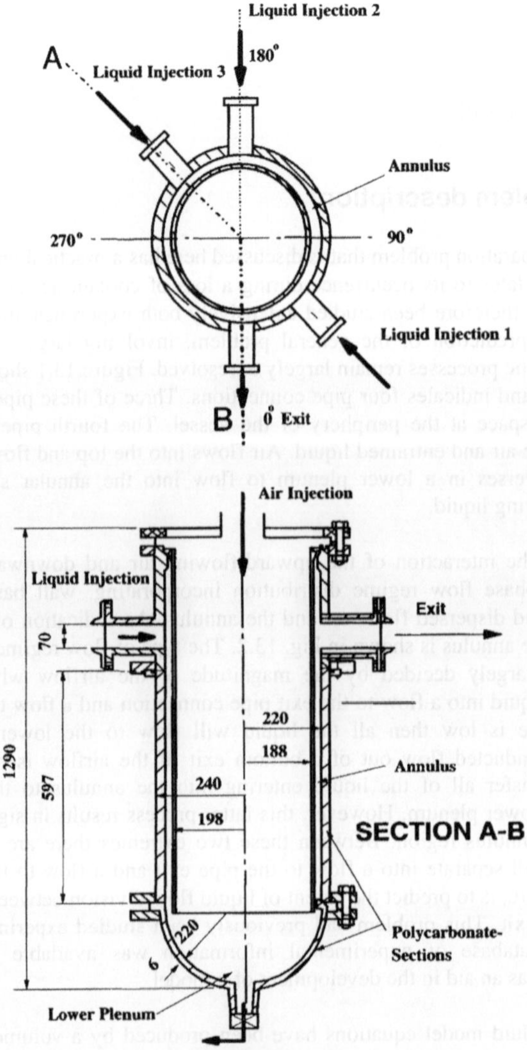

Fig. 13.1 Mixing vessel

13.4 Two fluid model – basic equations and closure relationships

13.4.1 Transport equations

The equations that are used in the computational exercise are set out below in full and consist of the continuity equations and momentum equations for the gas and liquid phases in each direction. The problem was modelled as a transient two-dimensional problem in the z–θ plane. The equations can therefore be presented in the Cartesian co-ordinate system.

$$\frac{\partial}{\partial t}(\alpha\rho_g)+\frac{\partial}{\partial x}(\alpha\rho_g u_g)+\frac{\partial}{\partial y}(\alpha\rho_g v_g)=0$$

$$\frac{\partial}{\partial t}((1-\alpha)\rho_l)+\frac{\partial}{\partial x}((1-\alpha)\rho_l u_l)+\frac{\partial}{\partial y}((1-\alpha)\rho_l v_l)=0$$

$$\frac{\partial}{\partial t}(\alpha\rho_g u_g)+\frac{\partial}{\partial x}(\alpha\rho_g u_g u_g)+\frac{\partial}{\partial y}(\alpha\rho_g u_g v_g)=-\alpha\frac{\partial p}{\partial x}+M_{ig_x}+M_{wg_x} \qquad (13.1)$$

$$\frac{\partial}{\partial t}(\alpha\rho_g u_g)+\frac{\partial}{\partial x}(\alpha\rho_g u_g v_g)+\frac{\partial}{\partial y}(\alpha\rho_g v_g v_g)=-\alpha\frac{\partial p}{\partial y}+\alpha\rho_g g+M_{ig_y}+M_{wg_y}$$

$$\frac{\partial}{\partial t}((1-\alpha)\rho_l u_l)+\frac{\partial}{\partial x}((1-\alpha)\rho_l u_l u_l)+\frac{\partial}{\partial y}((1-\alpha)\rho_l u_l v_l)=-(1-\alpha)\frac{\partial p}{\partial x}+M_{il_x}+M_{wl_x}$$

$$\frac{\partial}{\partial t}(\alpha\rho_l v_l)+\frac{\partial}{\partial x}((1-\alpha)\rho_l u_l v_l)+\frac{\partial}{\partial y}((1-\alpha)\rho_l v_l v_l)=-(1-\alpha)\frac{\partial p}{\partial y}+(1-\alpha)\rho_l g+M_{il_y}+M_{lw_y}$$

In the above equations, closure relationships have to be supplied for the interfacial drag terms, M_{ig} and M_{il} and the wall shear terms M_{Wl}, M_{Wg}. These terms are critical in determining the correct prediction of the flow processes. The amount of liquid that is bypassed out of the break is dictated by the drag forces exerted by the air flow and will vary depending on the local conditions in the annulus.

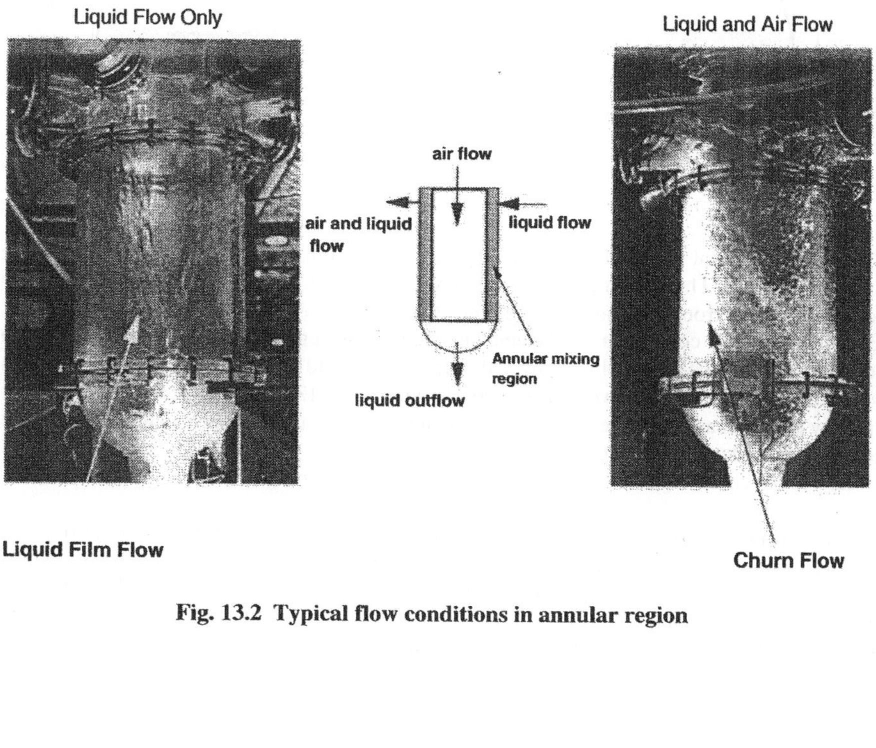

Fig. 13.2 Typical flow conditions in annular region

13.4.2 Model closure relationships

Flow Regime Map

To supply appropriate terms for M_{Wl}, M_{Wg}, M_{il} and M_{ig} for the particular two phase conditions that occur in the annulus, a flow regime map has been constructed from observations and physical intuition gained from the experimental study. Previous work, as discussed by Dempster and Abouhadra (**3**) has established that the exit flow characteristics are predominately independent of the inlet liquid flowrate. This feature has also been observed for the flow regimes, as indicated in the work of Abouhadra (**4**) and shown in Fig. 13.3. It can be seen in this figure that the flow regimes have similar distributions for approximately the same non-dimensional gas flowrate J^*g and for different liquid flow rates, J^*_W. These results suggest that it may be possible to create a flow regime map that is independent of the inlet liquid flowrate. The independence of the liquid flowrate has only been established using the gross features of the flow. To develop a model further it has been assumed that the local flow regime at any point in the annulus can be completely described by local conditions independent of the liquid flowrate.

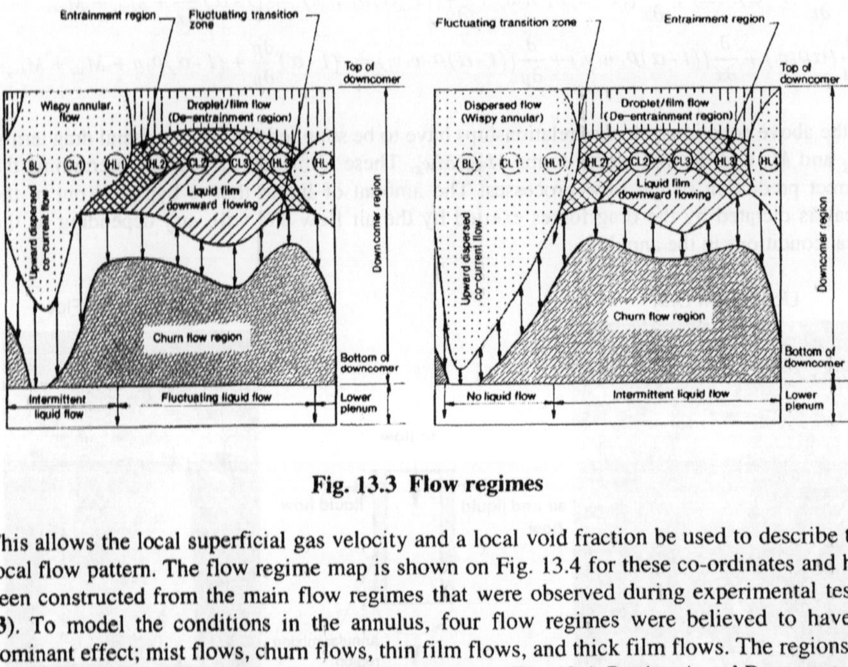

Fig. 13.3 Flow regimes

This allows the local superficial gas velocity and a local void fraction be used to describe the local flow pattern. The flow regime map is shown on Fig. 13.4 for these co-ordinates and has been constructed from the main flow regimes that were observed during experimental tests, (**3**). To model the conditions in the annulus, four flow regimes were believed to have a dominant effect; mist flows, churn flows, thin film flows, and thick film flows. The regions in which these regimes are believed to exist are shown on Fig. 13.4. Region A and B are parts of the flow map that are not expected to be encountered. To ensure that the calculations will not fail due to the lack of closure laws, Region A defaults to the mist regime and Region B defaults to the churn regime if the solution falls in these regions.

The mist flow regime accounts for all the dispersed flows that might occur during the flooding process and at this stage in the modelling process also accounts for regimes where the liquid may be partly wall based, i.e. film-mist type flows. The modelled churn flow regime describes a chaotic flow pattern observed in the annulus but only describes the conditions where a co-current flow of liquid and gas is expected. The modelled film flow regimes actually accounts for observed liquid film flows but also accounts for churn flows when the conditions lead to a downward flow of liquid. The film flow is further separated into a thin film and a thick film regime. For most cases a film flow condition will consist of the thin film regime. However, thick liquid films were observed local to the liquid injection positions due to the build up of water as the liquid slowed down to change from an inlet radial direction to a circumferential or axial direction. In addition, a stagnation point is likely to exist which may account for the liquid completely filling the downcomer gap in these regions at high inlet liquid flowrates. These conditions being related to entrance effects are not normally investigated in two phase flows and therefore would not be described by any constitutive drag relationship that represents thin films flow conditions. It was therefore necessary to separately model the two observed film flow regimes.

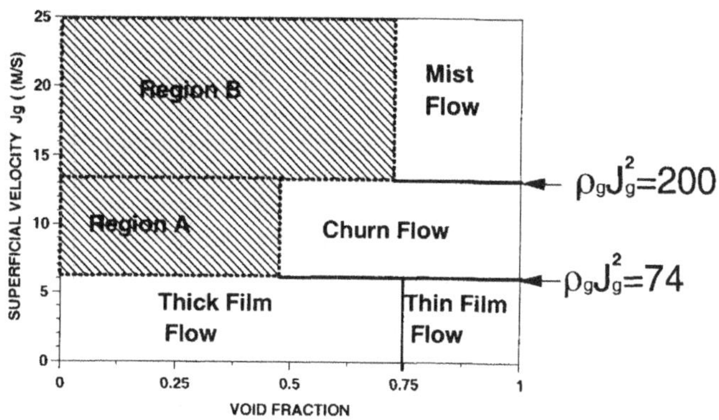

Fig. 13.4 Flow regime map

The main difficulty in constructing the map was in the identification of regime transition points. Two dominant transition points were the film/churn condition and the churn/mist condition. Both the film/churn and the churn/mist condition have been determined by identifying the transition superficial velocity from experimental observation for tests at approximately atmospheric conditions. Both regime transition points are presented in accordance with Hewitt and Roberts (5) flow regime map for vertically upwards air/water flows.

The film/churn transition condition was found to be

$$\rho_g J^2_g = 74 \tag{13.2}$$

and the churn/mist condition

$$\rho_g J^2_g = 200 \tag{13.3}$$

The film/churn transition was based on the observation that at this condition an upward flow of liquid was observed in the churn region. This identification is distinct from the initiation of churn conditions in the annulus since at lower gas flowrate churn conditions were observed with no noticeable change in the liquid downflow into the lower plenum. To account for this observation the churn flow regime was split into two sub-regions depending on the observed direction of the film flow, i.e. a churn regime in which the dominant liquid flow direction was downwards and another when the air flow was sufficient to produce significant upward liquid flow. The film/churn transition identified by equation 13.2 refers to a condition which separates the downward and upward liquid flow conditions. The churn/mist transition point, (equation 13.3) was found to be consistent with the Hewitt and Roberts (5) condition for churn to annular transition in tubes even though the current application has an annular geometry.

Closure relationships

The closure relationships that are required to produce a complete mathematical model of the flow phenomena do not exist for the specific counter-current flow conditions and geometry that are being investigated here. Therefore, this study has resorted to interfacial drag and wall friction correlations available from the literature, but applied judiciously in the context of the flow regime map. Table 13.1 shows the closure relationships. It should be noted that the terms M_w, M_{il} and M_{ig} are the forces per unit volume. When the differential equations are transformed by the Finite Volume Method to algebraic form the interfacial terms are presented directly as forces (N). Table 13.1 presents closure equations for the Finite Volume Form. For the interfacial momentum transfer, the model uses the Wallis (6) annular flow model in the liquid film flow regime, the Ishii and Chawla (7) relationship for the churn flow regime and droplet drag laws for the mist regime. (Clift *et al*, 1978).

13.5 Computational modelling

13.5.1 Geometric modelling

The numerical grid that was used for the majority of the calculations was created by 'unwrapping' the annular geometry of the downcomer into a slab geometry. The grid has uniformly distributed cells and consists of a 32 × 30 distribution (960 cells) where each cell size is approximately 4 × 4 cm. The boundary conditions therefore consist of inlet liquid cells representing the connecting pipe junctions and air inlet cells representing the lower plenum region whose connection with the core allows the inflow of air. In the experimental test facility the lower plenum region is connected to an outlet pipe that allows the penetrated liquid to flow to a measurement tank. However, in the numerical model this outlet flow boundary condition was found difficult to implement. Alternatively, the lower plenum was modelled as a large but closed volume allowing the liquid to build up as the transient simulation proceeded. The volume was made so large that the lower plenum inlet flowrate could not significantly affect the lower plenum liquid content during the calculation period. As noted above, the geometric model represents the annular region and lower plenum as a two dimensional Cartesian slab. This introduces a distortion of the lower plenum which may affect the predictions. However, the assumption is that the flow in the annular space is insensitive to the lower plenum conditions. This was consistent with observations during the experimental tests, except when the lower plenum was nearly full of liquid. For this condition, the entrainment of lower plenum liquid was apparent which seemed to have an influence on the liquid content in the lower annulus regions. However, during all the experimental tests the liquid level in the lower plenum was maintained at liquid levels that prevented such an occurrence.

13.5.2 Boundary conditions

Four main boundary conditions require to be implemented to simulate the experimental tests:

(i) Cyclic boundary conditions at x = 0 and x = W;
(ii) Air inlet mass flow at the lower plenum region;

(iii) Outlet pipe location;
(iv) Inlet water flowrate for the three connecting pipes.

Table 13.1 Interfacial drag and wall friction closure relationships

Flow Regime	Interfacial Drag	Wall Drag
Mist Flow	$F_i=C_i A_i \frac{1}{2}\rho_g V_r^2$ $Re_d<0.103 \quad C_i=240$ $0.103<Re_d<989.0 \quad C_i=\frac{24}{Re_d}(1.0+0.15 Re_d^{0.687})$ $Re_d>989.0 C_i=0.44$ $Re_d=\frac{D_d V_r \rho_g}{\mu_g} \quad W_d=\frac{D_d \rho_g V_r^2}{\sigma} \quad A_i=\frac{1.5(1-\alpha)}{D_d}V_{cell}$	$Fw_l=A_{wl} f_w \frac{1}{2}\rho_l V_l^2 \quad Fw_g=A_{wg} f_w \frac{1}{2}\rho_l V_l^2$ $fw=\frac{16}{Re_f} \quad Re_f \le 1000$ $fw=0.0791 Re_f^{-0.25} \quad Re_f>1000$ $\mu=\left[\frac{x}{\mu_g}+\frac{1-x}{\mu_f}\right]^{-1}$ $A_{W_g}=\alpha A_W \quad A_{W_L}=(1-\alpha)A_W \quad A_W=\frac{2V_{cell}}{S}$
Churn Flow	$F_i=C_i A_i \frac{1}{2}\rho_g V_d^2$ $V_d=\alpha V_r \quad C_i=\frac{8}{3}$ $A_i=\frac{1.5(1-\alpha)}{D_d}V_{cell} \quad W_d=\frac{\rho_g D_d V_d^2}{\sigma}=12$	$F_w=0$
Thin Film	$F_i=C_i A_i \frac{1}{2}\rho_g V_g^2$ $C_i=0.005(1+300\frac{\delta}{D})$ $\frac{\delta}{D}=\frac{\delta}{S} \quad \delta=\frac{S}{2}(1-\alpha) \quad A_i=\frac{2V_{cell}}{S}$	$F_w=Fw_l=A_w f_{wl} \frac{1}{2}\rho_l V_l^2$ $fw=\frac{16}{Re_f} \quad Re_f \le 1000$ $fw=0.0791 Re_f^{-0.25} \quad Re_f>1000$ $Re_f=\frac{4\delta\rho_L V_L}{\mu_L} \quad \delta=\frac{S(1-\alpha)}{2} \quad A_W=\frac{2V_{cell}}{S}$
Thick film	$F_i=C_i A_i \frac{1}{2}\rho_g V_g^2$ $C_i=0.005$ $\frac{\delta}{D}=\frac{\delta}{S} \quad \delta=\frac{S}{2}(1-\alpha) \quad A_i=\frac{V_{cell}}{S}$	$F_w=Fw_l=A_w f_{wl} \frac{1}{2}\rho_l V_l^2$ $fw=\frac{16}{Re_f} \quad Re_f \le 1000$ $fw=0.0791 Re_f^{-0.25} \quad Re_f>1000$ $Re_f=\frac{4\delta\rho_L V_L}{\mu_L} \quad \delta=\frac{S(1-\alpha)}{2} \quad A_W=\frac{2V_{cell}}{S}$

The cyclic boundary, the air inlet mass flow, and the exit pipe are specified in the normal way. However, the inlet liquid flowrate condition is a particularly difficult boundary condition to impose accurately. The connecting pipes are horizontal connections that do not run full during injection conditions therefore the water approaches in a stratified flow condition. Furthermore, the flow conditions that arise from liquid entry into the annulus and redistribution around the downcomer is a complex multi-dimensional flow problem in itself. The experimental tests suggested that the extent of liquid distribution around the annulus was connected with the flow conditions at the pipe annulus connection. Unfortunately the model in its present form could not address these issues. Therefore, a boundary condition was required at the inlet to account for the inlet mass flow and the redistribution effect. This was done by imposing an inlet mass flow corresponding to the measured inlet water flow and a circumferential (horizontal directions) momentum source to provide the necessary distribution effect. The inlet mass flow, M_{lin} for each cold leg was uniformly distributed over N inlet cells, i.e.,

$$M_{lin} = \sum_{i=1}^{N} M_{l_{cell}} \tag{13.4}$$

The circumferential momentum source term was modelled by the inlet velocity V_{lin} which took account of the stratified nature of the inlet flow, i.e.

$$Mom_l = C_o \dot{M}_{lin} V_{lin} \tag{13.5}$$

C_o is a correction factor and V_{lin} is determined from

$$V_{l_{in}} = \frac{\dot{M}_{l_{in}}}{\rho A_{cl}(1-\alpha)_{in}} \tag{13.6}$$

The void fraction at the inlet, α_{in} was determined experimentally by measuring the circumference of the outside diameter of the injection pipe wetted by inlet flowing water. This was converted to a void fraction. A linear regression established the following relationship between void fraction at the inlet section to the downcomer as a function of the inlet mass flowrate, i.e.

$$\alpha = -0.219 M_{lin} + 1.0 \tag{13.7}$$

The correction factor C_o was found by comparison of calculated and experimental tests to achieve a good correspondence between the observed liquid distribution at the bottom of the annulus and the calculated distribution, for zero airflow conditions. It was found that a value of C_o=2 achieved the desired results.

13.5.3 Calculation procedures

Initial investigations were carried out on a relatively coarse mesh grid but modified to the 30×32 for all experimental test simulations. The numerical scheme is based on the Interphase Slip Algorithm (IPSA) and is a modified procedure of the well known SIMPLE algorithm. A hybrid differencing scheme is used for the convection terms and a fully implicit scheme is used for the temporal discretization. A time step of 0.002 seconds was used and is slightly above the Courant Friedrichs Lewy (CFL) stability limit for an explicit calculation that was found to be approximately 0.0015 seconds. This value was found to be satisfactory for all calculations carried out in this study. The convergence criteria was set at 1 per cent however, no noticeable change in the mean results were found when tested at 0.1 per cent and at a time step of 0.001 seconds. The calculations were performed on a Gateway 2000 Pentium P5-90 PC. The simulations were carried out for period from 3 to 8 seconds real-time and usually started from the final time dump of a previous simulation. A calculation over a period of 8 seconds real-time required approximately 15 hours of cpu time.

13.6 Computational results

13.6.1 Overall comparison

The simulations of each individual experiment allowed the liquid exit flow to be calculated directly from the exit boundary cell. The predictions of the liquid flowrate are compared with experimentally measured values and are shown graphically on Fig. 13.5. In this graph it can be seen that the predictions are generally good for the high-exit liquid flowrate conditions and clearly within a ± 25 per cent range.

For low exit flow conditions a poorer accuracy was noted. This is associated with the difficulty in modelling the entrainment conditions local to the break.

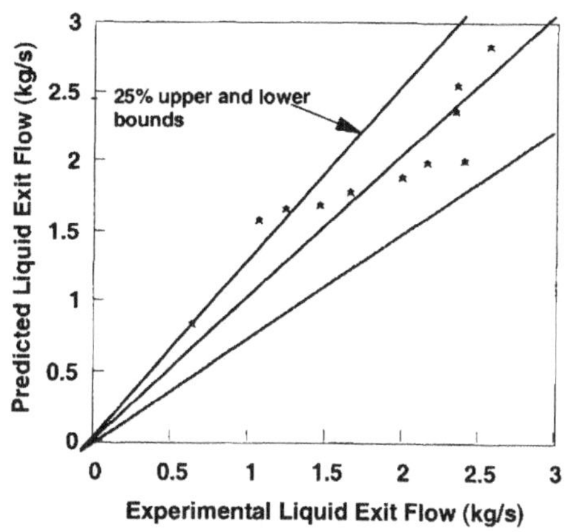

Fig. 13.5 Comparison of experimental and predicted liquid exit flow

To indicate how the mathematical model performs at a detailed level a high liquid flow bypass test was chosen where 70 per cent of the injected liquid was bypassed out of the exit. The calculated exit flow over a period of the transient is compared with the experimental values on Fig. 13.6. The predicted results fluctuate about an average value which slightly underpredicts the experimental results. Figures 13.7, 13.8, and 13.9 shows the air velocity vector, liquid velocity vector and liquid fraction distribution, respectively at the end of the calculation period. The air velocity vectors, shown on Fig. 13.7, indicate that the distribution is mainly uniform in the lower annular region. The liquid fraction distribution is predicted to be more non-uniform with fluctuating regions of high liquid content. The liquid velocity distribution is shown in Fig. 13.8 and indicates the flow paths of the injected liquid.

The liquid entering from injection pipe 1 flows directly to the break. The liquid entering from injection pipes 2 and 3 distributes itself into the annulus with the majority flowing into high regions of air velocity and subsequently being swept out of the break. The remaining liquid, flows to the lower plenum in regions directly under the break.

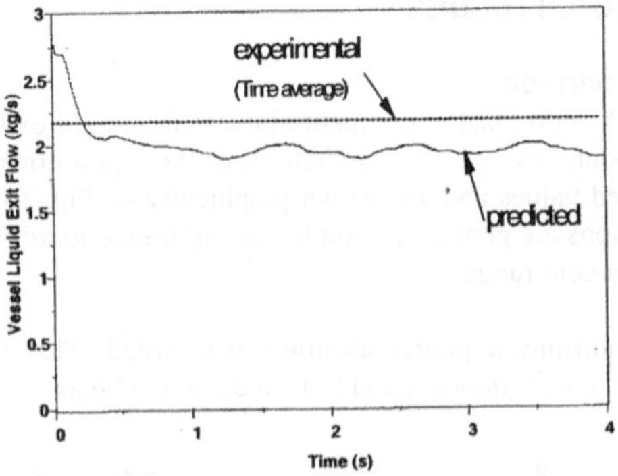

Fig. 13.6 Exit flow results

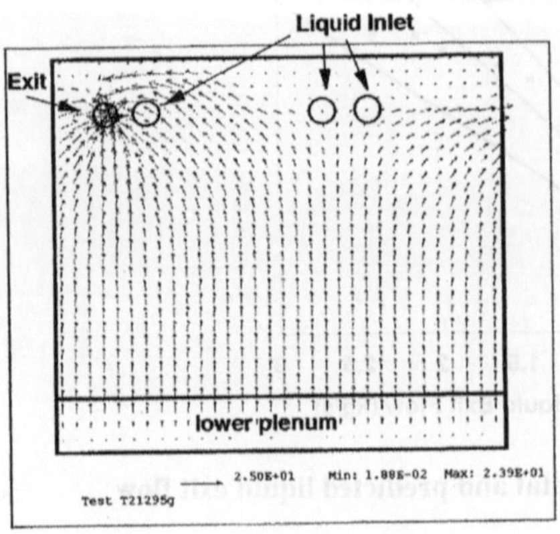

Fig. 13.7 Typical air flowrate

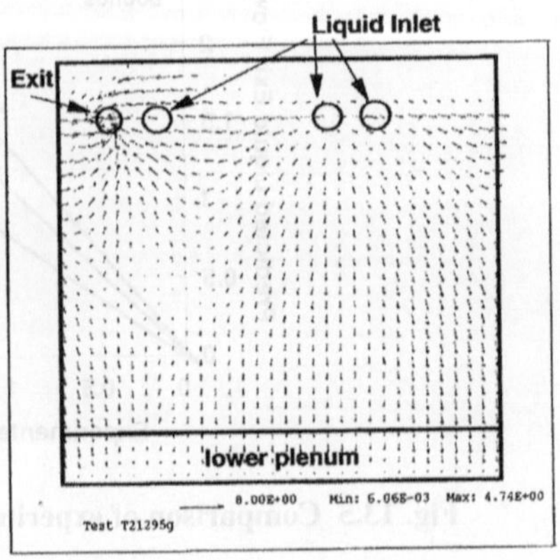

Fig. 13.8 Typical liquid flowrate

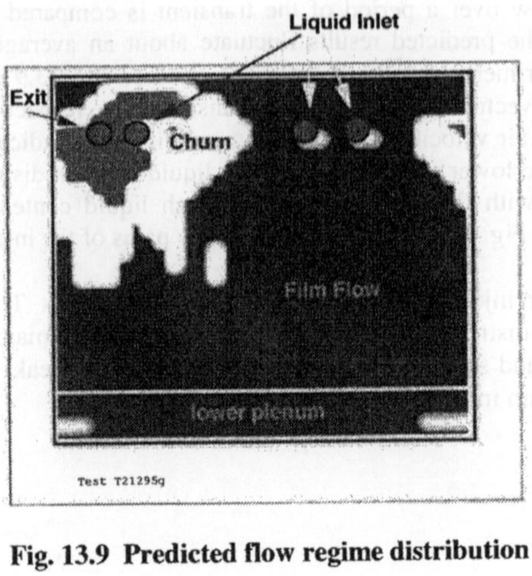

Fig. 13.9 Predicted flow regime distribution

The prediction of the flow regime distribution is shown on Fig. 13.9 and can be compared to experimentally similar conditions in Fig. 13.3(a). The predicted regime distribution is satisfactory but must be interpreted in the context of the modelled regime definitions. The modelled mist and churn regimes attempt to account for co-current liquid flow conditions, while the modelled film regime attempts to account for downward flow of liquid which could exist in the observed film and churn regimes. Accounting for these definitions, Fig. 13.9 indicates that predicted and observed flow regimes correspond reasonably well.

13.7 Discussion

From the computational results it is apparent that the model does a reasonable job of predicting the separation of liquid mass flowrates. It has also been established that this predictive capability continues for a range of different liquid injection flowrates and pressure conditions, and provides some confidence in the model. However, the question originally posed regarding the general applicability of the model requires still to be addressed. The closure of the model was obtained by partly using generally available relationships and information gained from model testing. Table 13.2 summarizes the extent of the model closure relationships.

Table 13.2 Model closure conditions

Closure Condition	Source
Interfacial Transfer law	Available from literature
Wall Shear laws	Available from literature
Flow Structure	Derived from experimental model testing
Boundary Conditions	Derived from experimental model testing

The apparent success of the prediction relies heavily on the availability of interfacial momentum transfer relationships. The relationships used in this study were initially developed for pipe flows where one-dimensional approximations have found to be successful. The general applicability of constitutive relationships to multi-dimensional flows was never assured, however, the reasonable correspondence between predictions and experiments give some confidence in the approach. Unfortunately, the correct application of the relationships required prior knowledge of the flow structure (the flow regime map) and at present these conditions could only be accurately obtained from experimental testing .

13.7 Conclusions

A mathematical model has been constructed from a two fluid representation of the two-phase flows that occur during a mixing/separation process in an annular geometry. The model has been based on the observed conditions that take place in an experimental air water facility and tested against the experimental measurements. The following conclusions can be made from the study:

(i) The model satisfactorily predicts the exit and lower plenum flowrates when compared with the experimental data;

(ii) The generally applicability of such models for two phase flow problems is far from guaranteed and still largely depends on experimentation.

13.9 References

(1) **Hestroni, G.** and **Yadigaroglu, G.** (1992) Multiphase flow and heat transfer: bases, modelling and applications in (A): The nuclear power industries, (B): The process industries, Short Course Notes, Zurich, Switzerland, March.

(2) **Ishii, M.** and **Mishima, K.** (1984) Two fluid model and hydrodynamic constitutive relations, *Nuclear Engng Des.*, **82**, 107–126.

(3) **Dempster, W.** and **Abouhadra, D.** (1994) Multi-dimensional two-phase flow regime distribution in a PWR downcomer during an LBLOCA refill phase, *Nuclear Engng Des.*, **149**, 153–166.

(4) **Abouhadra, D.** (1992) *Two phase flow regimes in a PWR vessel downcomer during the refill phase of a large LOCA*, MPhil thesis, University of Strathclyde, Glasgow.

(5) **Hewitt, G.** and **Roberts, D. N.** (1969) Studies of two-phase gas-liquid flow patterns by simultaneous X-ray and flash photography, AERE-M 2159, HMSO.

(6) **Wallis, G. B.** (1969) *One Dimensional Two-phase Flow*, McGraw Hill Book Co., New York.

(7) **Ishii, M.** and **Chawla, T. C.** (1979) Local drag laws in dispersed two-phase flow, NUREG/CR-1230.

(8) **Bharathan, D., Wallis, G. B.,** and **Richter, H. J.** (1979) Air water counter current annular flow, EPRI Topical Report, NP-1165.

14

CFD Computation of Air–oil Separation in an Engine Breather

I Care, C Eastwick, S Hibberd, K Simmons, and **Y Wang**

Synopsis

In high-performance engines an oil–air mixture is formed in transmission chambers by a combination of air leakage flow past seals and oil injection. To maintain engine reliability and reduce oil pollution to the environment the mixture must be separated before the oil can be reused and excess air expelled. Typically a breather is used which comprises a rapidly rotating cylindrical housing filled with a porous separator matrix. The breather is designed to exhaust the air from the centre whilst the oil centrifuges outwards to be scavenged from the enclosing stationary breather chamber.

CFD computations have been developed based on the commercial code CFX4. Initially, a simplified model for the highly rotational breather flow was created, followed by a geometrically accurate model of an industrial breather system. The turbulent airflow inside the breather chamber was calculated and oil droplet motion was studied using Lagrangian particle tracking. Droplets below a critical diameter are found to impinge on the porous matrix; otherwise they impinge differentially on the stationary chamber walls. Performance of the air–oil separation is predicted to vary significantly with shaft speed and breather configuration.

14.1 Introduction

Within the transmission system of aero or other high-performance engines, breathers are used as separation devices for weak oil/air mixtures. Oil is injected for lubrication and cooling and airflow is also present from positive pressure sealing or leakage through bearings and seals. Both the retention of oil within the transmission system and minimizing the environmental impact of exhausted oil droplets are important factors. The aim of the separation device is to maintain the internal pressure balance in the transmission chamber and recover oil from the oil/air mixture that would otherwise be vented with the air to the atmosphere. It is, therefore, important to understand and quantify the physical phenomena relating to the separation process.

A common design for a breather consists of a housing, containing a porous metal (or plastic) matrix, rigidly mounted on a rapidly rotating shaft within a stationary chamber. Figure 14.1 shows a schematic diagram of the key components of a typical breather configuration and identifies a tangential air/oil inlet and separate oil scavenge from the outer chamber surface. The rotating components generate a flow configuration within the chamber, and also provide a vented outlet for flow (predominantly air-phase) passing radially inwards through the porous material and exhausting through a duct in the centre of the shaft. In practise, the breather may be a self-contained unit attached to the engine or located as part of a main shaft. The speed of the shaft typically varies between 10 000 and 20 000 r/min and imparts significant outward radial motion to oil collected within the separator matrix.

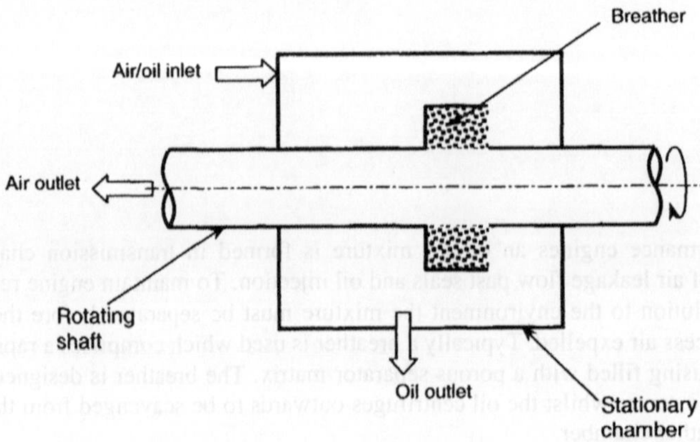

Fig. 14.1 Schematic diagram of a typical breather with rear face open

An example of a suitable breather matrix material is Retimet, a patented metal foam material developed by Dunlop Equipment (1), (2). Figure 14.2 shows a micrograph of a sample of Retimet that suggests it may act dynamically as a highly porous medium. In this study flow through the separator matrix is taken to conform to a Darcy flow law (differential pressure directly proportional to velocity).

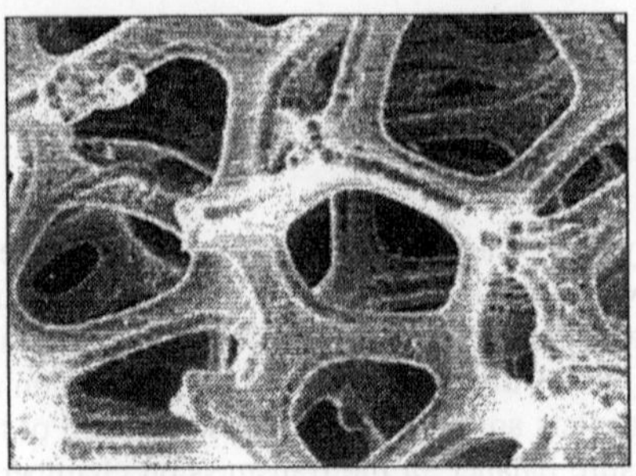

Fig. 14.2 Micrograph showing the structure of Retimet

The separation of the oil/air mixture occurs by two main mechanisms. The first is a centrifuging action where oil droplets are 'spun out' by the internal air flow onto the walls of the chamber where they subsequently form a film that is collected and scavenged from the breather chamber. The second is for a mixture of air and oil droplets to enter the separator matrix of the breather. Inside the matrix most of the oil drops coalesce and are centrifuged to the walls of the housing where it passes through a series of holes back to the chamber. Simultaneously the air, with any residual oil is vented through the centre of the hollow shaft.

The related physical phenomena involving the oil/air separation process within a rapidly-rotating (annular) flow are complicated and relatively few studies have been reported due to the difficulty of undertaking both experimental measurements and computations. Currently, breather manufacturers use performance models derived from data compiled from in-house testing but these provide no information about detailed behaviour within the chamber itself. Wittig *et al.* **(3)**, **(4)** and Glahn *et al.* **(5)–(7)** have reported work on aspects of the physical phenomena occurring in aero-engine bearing chambers which involve the flow between an outer stationary housing and a rotating shaft. Simple models for the oil film flow exist such as that presented by Chew **(8)** for the oil film on the housing of a bearing chamber using an integral approach, which gave good qualitative and quantitative agreement with measurements existing within the literature. Although these studies have provided valuable modelling information on the physical phenomena occurring inside aero-engine bearing chambers, they are not able to provide quantitative information relevant to breathers, due to the complex geometrical configurations and high speed of the rotating elements.

The aim of this study was to provide detailed insight into the behaviour of oil droplets released into the induced turbulent airflow generated within a specific breather chamber geometry (Techspace Aero, Fig. 14.3) for increasing shaft speed using the commercial CFD package CFX4.

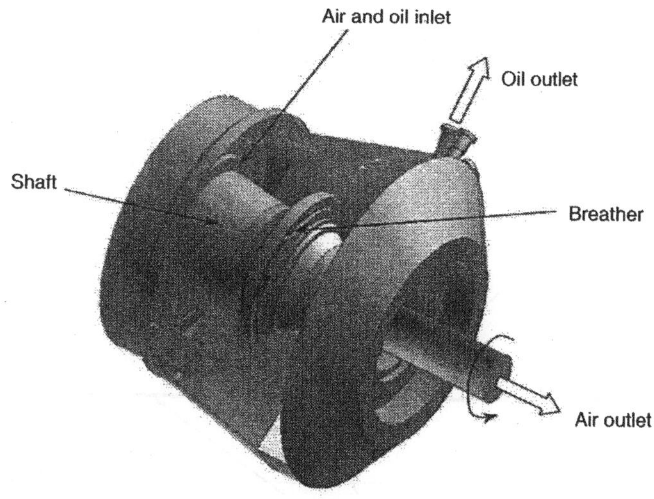

Fig. 14.3 Computational solid model of an aeroengine breather system

14.2 Methodology

The solution methodology adopted was to complete the calculations in two stages. In the first stage, a simplified axisymmetric model of the chamber was constructed with either the front face of the breather housing open or the rear face (see Fig. 14.1, where the rear face is open). Turbulent air phase computations were converged using CFX4 and oil droplets of specified sizes were tracked through the domain using Lagrangian particle tracking as a post-process (droplet momentum does not modify the flow field). This stage of the investigation highlighted potential computational problem areas and suggested the way the flow inside the more complex breather chamber would behave. The results from this preliminary stage were found to be useful in identifying the flow features inside the breather chamber and enabled a more informed implementation of the CFD model.

In the second stage, a geometrically accurate axisymmetric CFD model of the aeroengine breather system was constructed. Using a computational solid model of the breather ensured geometrical accuracy of the CFD model. Guided by results from the simplified model an appropriate grid density was chosen and turbulent air phase calculations of the flow were converged. Typically the droplet-laden airflow through the breather chamber is of sufficiently low concentration for the motion to be simulated by the Eulerian-Lagrangian approach. Oil droplets of a relevant size distribution were tracked through the computational domain using Lagrangian particle tracking.

14.3 Simplified axisymmetric model

The geometry and boundary conditions for the simple axisymmetric model are shown in Fig. 14.4. Air enters the chamber through an azimuthal slot representing the outer race of the shaft bearing. Physically air and oil enter the aeroengine breather chamber through this bearing. Air leaves the chamber through the centre of the shaft, which is modelled as a mass flow boundary. The shaft and the breather housing are rotated at an appropriate angular velocity but the breather itself remaining stationary.

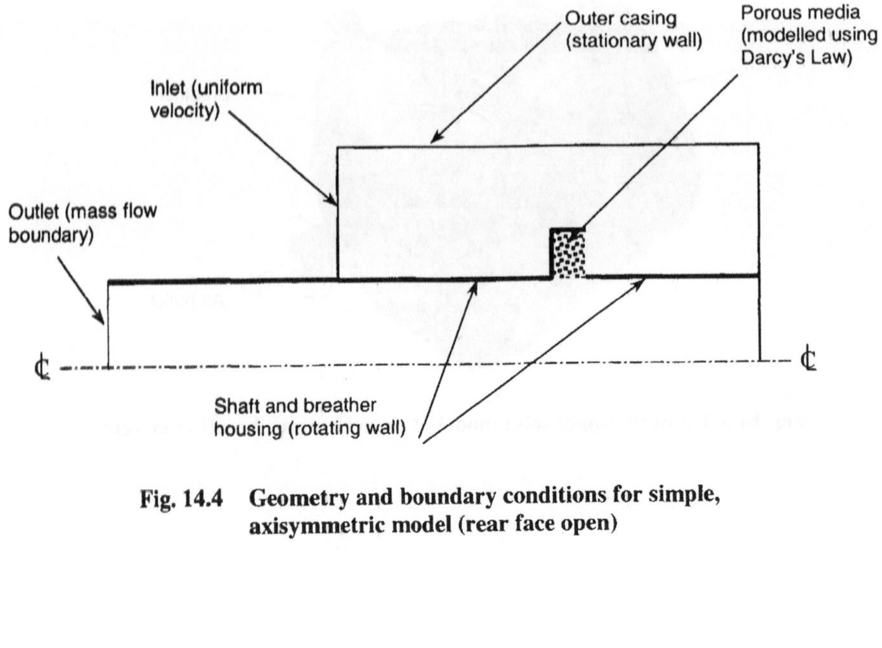

Fig. 14.4 Geometry and boundary conditions for simple, axisymmetric model (rear face open)

Values for the flow parameters are set at typical operating conditions with the air flux entering the breather chamber fixed at 0.05 kg/s and shaft speeds up to 13 000 r/min. The viscosity and the density of air are set to be 1.8×10^{-5} kg/m s and 1.2 kg/m^3 respectively.

The turbulent airflow inside the breather chamber is assumed to be axisymmetric and the standard k–ε turbulence model, as proposed by Launder and Spalding (**9**) and implemented in CFX, is used for the calculation. The Retimet inside the breather is modelled using a linear Darcy's law with porosity and flow resistance (inverse of matrix permeability) parameters set to be 0.95 and 6000 kg/m^{3}s respectively. For computational purposes, the air outlet is extended to include a sufficiently long outlet pipe, and conditions of fully-developed flow are imposed to ensure that results are not affected by the local prescription of the outlet-boundary condition. No allowance is made at this stage of the aircraft Mach number on the discharge of the outlet pipe. For clarity, the outlet pipe is not shown in the figures. On the outer casing and on the shaft it is assumed that any oil film is sufficiently thin not to affect the imposition of standard no-slip conditions; scavenge is also not modelled. Computations were performed on a grid with a total of 16 000 cells using a Silicon Graphics Origin 2000 workstation and suitably checked for grid independence.

Calculation of droplet motion was carried out using a Lagrangian tracking method, on the implicit simplifications that oil droplets can be modelled as hard spherical particles, collisions between particles can be ignored and that turbulent dispersion is negligible. Details of the Lagrangian method can be found in the CFX manual (**10**). Based on these assumptions and the computed air flow field, a representative droplet in the diameter range 1–600 μm is released at the centre of the inlet with an initial velocity prescribed at the local mean air flow velocity. The droplet trajectory calculation continues until the droplet either enters the open face of the breather or collides with a wall (zero co-efficient of restitution).

Plotted in Fig. 14.5 are the velocity vectors for the case of a rear-open breather housing, on an azimuthal plane for a shaft speed Ω = 11 000 r/min. It can be seen that the main air stream is directed towards the shaft, flows along the breather face and separates at the corner, moving outwards after the breather. This is probably due to the high shear stress generated from the rotating breather housing. Comparison with the airfield at lower shaft speed reveals that the direction change of the air stream is greater at the higher shaft speed. In addition, the air stream makes a further large change of direction having passed the breather housing, turning to enter the open face of the breather.

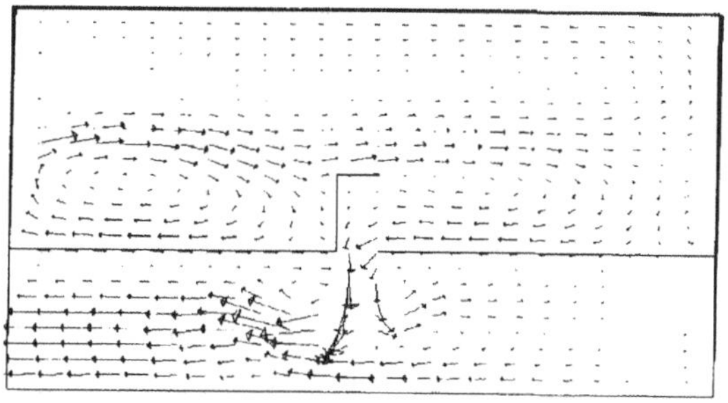

Fig. 14.5 Velocity vectors for the simplified geometry, rear face open, Ω = 11 000 r/min

For comparison, Fig. 14.6 shows the velocity vectors for the case of a front-open breather housing. As with the rear-open case the main air stream is directed towards the shaft and the upstream flow pattern is very similar for both cases.

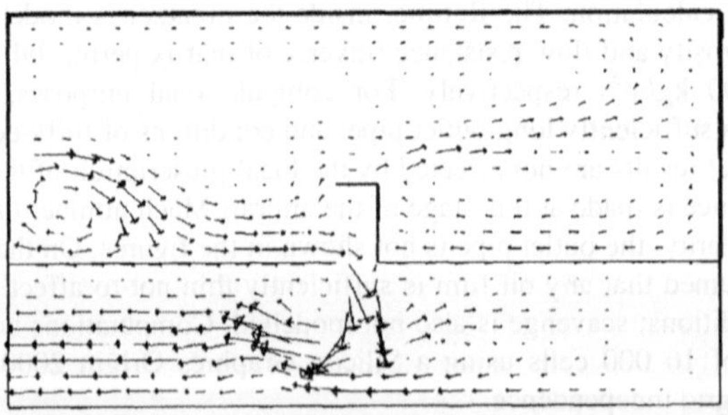

Fig. 14.6 Velocity vectors for the simplified geometry, front face open, Ω = 11 000 r/min

Shown in Fig. 14.7 is the projection of droplet trajectories onto an azimuthal plane for the rear-open breather with a shaft speed Ω = 11 000 r/min. As can be expected, the smaller droplets closely follow the air streamlines and enter the open face of the breather. As the size of the droplet diameter is increased, this behaviour changes due to the increasing importance of the inertia of droplets and, as a consequence, these droplets will no longer enter the open face of the breather. It is noted that droplets will be affected by the rotational velocity of the flow and will move outward towards the stationary outer casing. For very large droplets, the influence of the airflow to affect their initial inertia is small and these droplets will travel with their initial velocity until reaching the wall of the breather chamber.

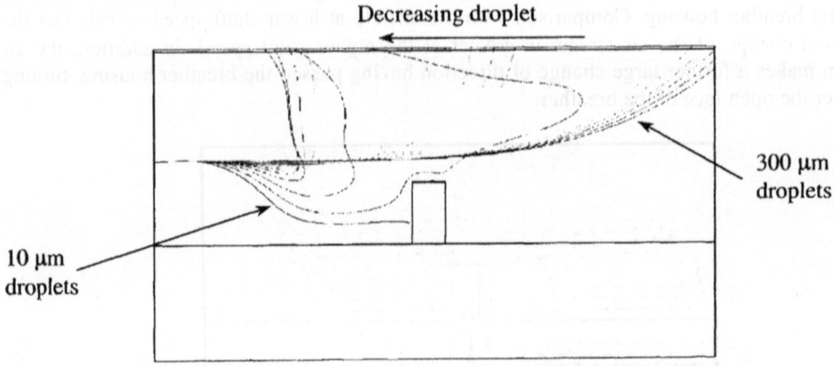

Fig. 14.7 Droplet trajectories, front face open, Ω = 5200 r/min

Figure 14.7 shows that larger droplets are centrifuged out to the walls of the chamber while smaller droplets impinge on the Retimet. It is therefore informative to define a critical droplet diameter, the maximum diameter for which droplets closely follow the airflow to reach the open face of the breather. This is illustrated in Fig. 14.8(a), where the critical droplet diameter is plotted as a function of shaft speed for both front-open and rear-open cases. It is seen that the critical droplet diameter decreases as the shaft speed increases for both cases and that this droplet diameter is significantly smaller in the rear-open case. This is consistent with the conjecture that in the rear-open case there is more streamline curvature such that slight deviation from the airflow will lead to non-impingement on the porous material. It is usual to relate the ability of particles to follow curving streamlines to the appropriate Stokes number. In this case there is no obvious appropriate Stokes number because the streamline curvature is in the axial direction while the predominant flow is azimuthal. Further work will be required to identify appropriate parameters. The results displayed in Fig. 14.8(a) suggest that at all shaft speeds droplets impinge on the Retimet although at high speed fewer droplets will impinge in the rear-open case compared with the front-open for the same approach droplet size distribution.

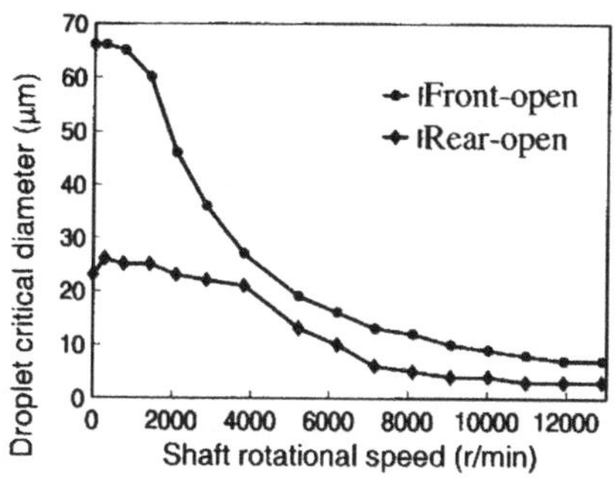

Fig. 14.8(a) Critical droplet diameter as a function of shaft speed for front-open and rear-open cases

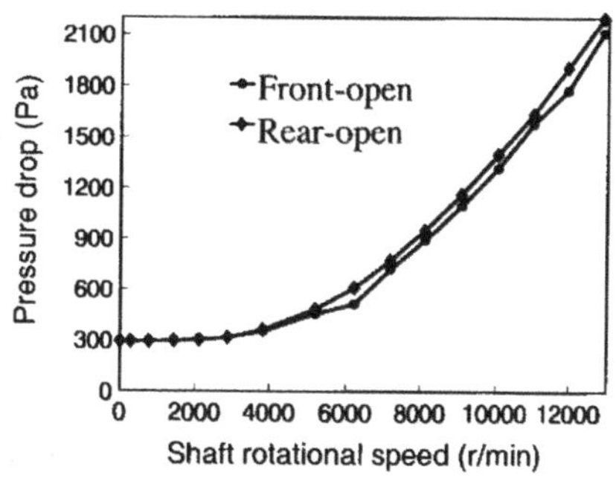

Fig. 14.8(b) Differential pressure between inlet and outlet as a function of shaft speed for front-open and rear-open cases

These results suggest that to improve separation, the shaft speed of the breather should be increased. This is not possible in the case of a direct shaft-mounted breather. For an auxiliary-driven breather rotation speed is restricted due to considerations of containment and burst strength. Increasing the speed also increases the pressure drop across the Retimet as shown in Fig. 14.8(b). If this pressure drop is too high, the bearing chamber sealing system will have a high back pressure, which could lead to inadequate bearing chamber sealing performance. Consequently, an optimal design will be a balance between the shaft speed and the configuration of the separator and bearing housing.

14.4 Model of an industrial breather system

The aeroengine breather system is shown as a solid model in Fig. 14.3 and as a diagrammatic cross-section in Fig. 14.9. Two bearings (front and rear) support an internal rotating shaft and both bearings are lubricated by oil. The breather is mounted on the rotating shaft and in this study the rear face was open. The inside of the breather housing has slots that interface with holes in the shaft to allow the air/oil mixture to exhaust through the shaft. The housing contains blocks of matrix material. Air enters the bearing chamber via the front bearing housing (see Fig. 14.9). The air flow is divided into two parts, one part goes through the holes on a deflection plate which is located before the front bearing and is designated as to 'Inlet 1' and the second part enters through the front bearing and is labelled as 'Inlet 2'. This latter flow entrains oil droplets with the air as it enters the chamber. In a similar manner, oil droplets may also be introduced into the bearing chamber with the air flow being blown from the rear bearing. In addition, an oil film may be formed on the inner surface of the bearing chamber due to overflow of oil from the bearings. Since the front bearing is the larger of the two bearings, and the mass flow of oil and air associated with this bearing is an order of magnitude larger than that from the rear bearing, the rear bearing has not been modelled in the present work.

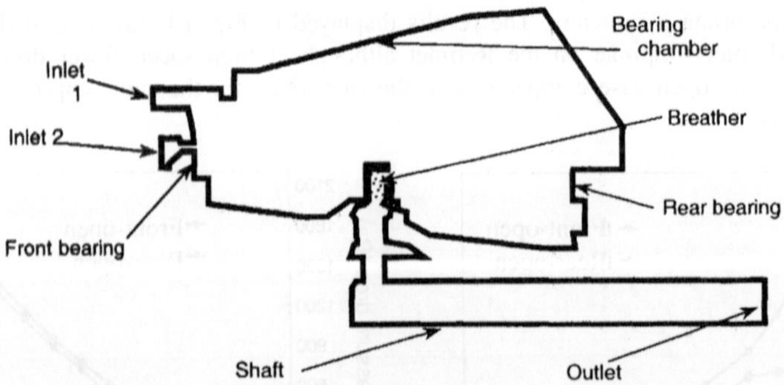

Fig. 14.9 Diagrammatic cross-section of the aeroengine breather, showing inlets and outlets

As with the simplified breather, the single-phase turbulent air flow in the bearing chamber and breather system is calculated on the assumption that the droplet-laden air flow through the bearing chamber is of sufficiently low concentration such that the existence of the oil phase does not affect the air flow significantly. The air flux, which enters through the front bearing housing through the two inlets to the bearing chamber, is varied within the calculations between 0.025 kg/s during a normal flight state and 0.05 kg/s at take off state. The air flux is divided into two parts through the two inlets with a ratio of 0.06 and the shaft rotates at 5200 r/min. A comparison of this shaft speed and higher shaft speeds is made.

Uniform inlet velocities at the two inlets to the bearing chamber were determined from the air flux and the ratio of the air flux going through the front bearing. An outlet boundary condition (mass flow boundary) was prescribed at the exit of the shaft, a stationary wall boundary condition was given for the outer stationary casing and a rotating boundary condition was set for the shaft and breather housing. The grid, illustrated in Fig. 14.10, is a compromise between the capability of the computer and the size of the first cell required by using the wall function in the standard k–ε turbulence model. The total number of cells is 19 000.

As with the simplified breather, droplet trajectories were calculated using the Lagrangian approach and based on the assumption that droplet concentration is sufficiently low that the air flow field is not influenced by the droplet motion. Turbulent particle dispersion was also not considered. Droplet trajectories were calculated for five shaft speeds: 1, 3000, 6000, 9000, and 12 000 r/min and two air fluxes: 0.025 kg/s and 0.05 kg/s.

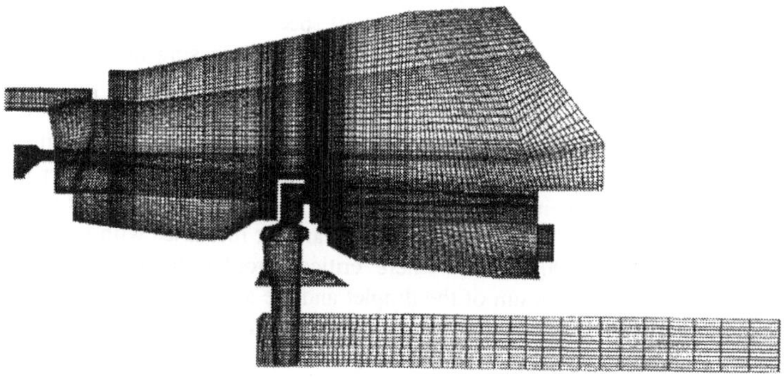

Fig. 14.10 Axisymmetric grid used to model the aeroengine breather (19 000 cells)

Figure 14.11 shows velocity vectors inside the breather chamber for an inlet air flux of 0.05 kg/s and a shaft speed of 6000 r/min. The large recirculation upstream of the breather housing is the most notable flow feature although a weaker recirculation also exists between the breather and the outer wall of the casing. Figure 14.12 shows velocity vectors for the same inlet air flux and a rotational speed of 12 000 r/min. Comparison shows that with the higher rotation speed the flow impinging on the breather housing is deflected outwards causing the recirculation between breather and housing to move towards the inlet.

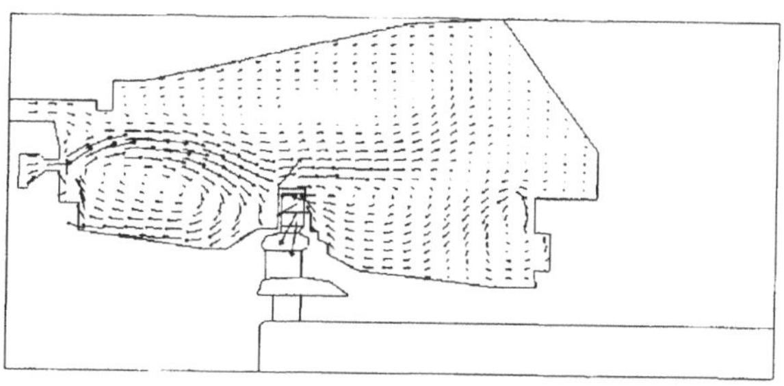

Fig. 14.11 Velocity vectors for an inlet air flux of 0.05 kg/s and a shaft speed of 6000 r/min

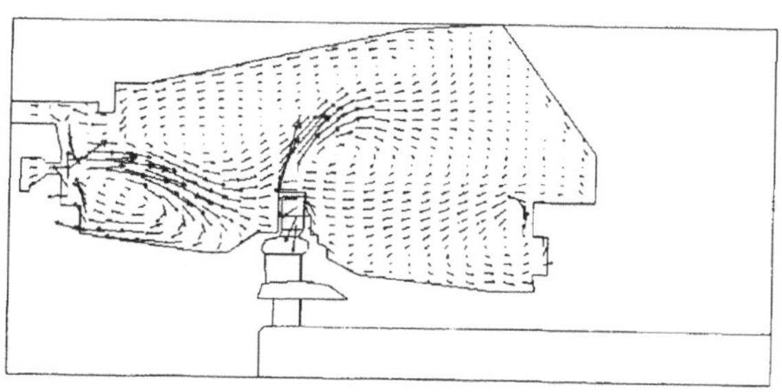

Fig. 14.12 Velocity vectors for an inlet air flux of 0.05 kg/s and a shaft speed of 12 000 r/min

In Fig. 14.13 droplet trajectories for droplets in the size range 0 to 600 μm are shown for the case where the inlet air flux is 0.025 kg/s and the shaft speed is 6000 r/min. Only droplets of less than 100 μm are not centrifuged directly to the stationary casing and only very small droplets impinge on the Retimet.

As with the simplified breather, a critical diameter can be identified above which the droplets will not impinge on the Retimet. A graph showing the variation of critical diameter with shaft speed for the two inlet air fluxes is displayed in Fig. 14.14. The graph contains two regions, 0 to approximately 1500 r/min, where critical droplet diameter increases with increasing shaft speed, and 1500 r/min to 12 000 r/min, where critical droplet diameter decreases with increasing shaft speed. The momentum of the droplet and the amount of streamline curvature are the two main factors affecting whether or not a droplet will follow the airflow. The fixed air flux corresponds to a fixed droplet momentum and so Fig. 14.14 suggests that at low rotation speed the amount of streamline curvature between inlet and the open face of the Retimet actually decreases. At higher speeds, comparison of Figs 14.11 and 14.12 shows that an increase of speed increases the angle at which the inlet jet is deflected from the breather housing, increasing streamline curvature. Consequently only smaller droplets will be able to follow the air to the Retimet.

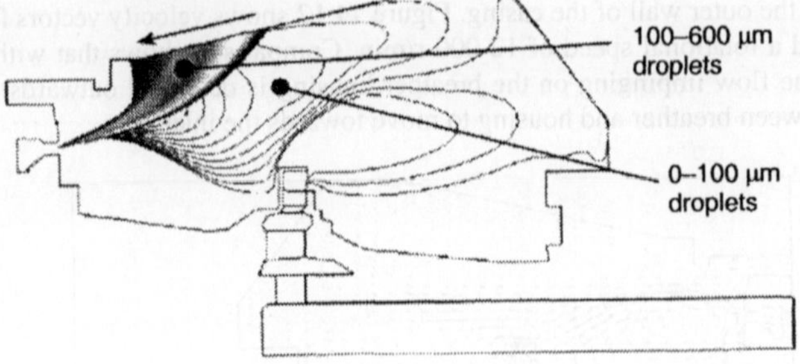

Fig. 14.13 Projection of trajectories for droplets in the size range 0 to 600 μm, inlet air flux 0.025 kg/s and shaft rotation speed 6000 r/min

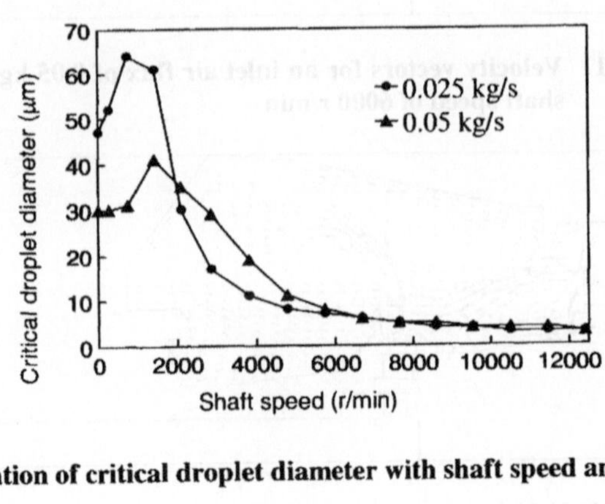

Fig. 14.14 Variation of critical droplet diameter with shaft speed and inlet air flux

14.5 Industrial application

The initial findings from the simplified axisymmetric model have led to a greater understanding of the flow processes operating within a breather chamber. In particular the results from this model indicated that the significance of the exit path of the air from the Retimet had been overlooked. An immediate result is a set of redesigned breathers for testing. Rig testing of the design improvements suggested by this simplified model have led to an improvement that will be evaluated on a development test engine shortly.

It is expected that further CFD development of the model for the complex breather chamber will lead to a revision of the design criteria for oil/air separators and their operating conditions. It is anticipated that even more efficient separation will be required in future, arising from environmental considerations. If this requires a higher rotational speed then this will create larger pressure differentials and this in turn may have an impact on the type of bearing chamber sealing systems that are employed. For applications where the shaft speed is fixed by other factors it will be necessary to give careful consideration to the air and oil paths into and through the Retimet.

Initial results from LDA/PIV measurements on an aeroengine are consistent with the CFD model and confirm that the oil droplets that emerge from the Retimet are between 1 μm and 14 μm.

14.6 Conclusion

In this study the operation of an industrial breather design was investigated using CFD. The approach adopted was to perform preliminary work on a simplified model representing many of the geometrical and physical features of the actual component. Further work on a geometrically accurate model was then completed. This methodology was found to yield useful information even during the early stages of the study and considerably reduced the amount of effort required for the full model.

The results of the study indicate a number of issues relating to the design of breather systems. It is particularly noted that systems that induce air streamline curvature will enhance the separation process, encouraging most of the oil to be centrifuged to the walls of the chamber and separated prior to the Retimet itself. A rear-open breather housing is therefore potentially a better design than front open. It is also clear that in an efficient design only very small droplets are likely to enter the Retimet, as these are the ones that can follow the airflow. Increasing the rotation speed also appears to enhance separation for the breather investigated.

Ongoing work includes extensions to the study of the Techspace Aero breather and modelling of breather configurations as developed and used by Dunlop Equipment Limited.

14.7 Acknowledgements

The authors would like to acknowledge the financial support of the Commission of the European Communities under contract number BRPR-CT97-0597 and the assistance of the other partners in this package: Dunlop Equipment Limited and Techspace Aero.

14.8 References

(1) **Dunlop,** Retimet Data Sheet. Dunlop Equipment Limited, 1980.

(2) **Dunlop,** (1997) Dunlop Equipment Retimet deoilers for aero-engines, *Aircraft Eng. Aerospace Tech.*, **69**, 64–66.

(3) CFX User Guide, (1997) Release 4.2. AEA Technology, England.

(4) **Wittig, S., Glahn, A.,** and **Himmelsbach, J.** (1993) Influence of high rotational speeds on heat transfer and oil film thickness in aero engine bearing chambers, *ASME paper* 93-GT-209.

(5) **Wittig, S., Glahn, A.,** and **Himmelsbach, J.** (1994) Influence of high rotational speeds on heat transfer and oil film thickness in aero engine bearing chambers, *J. Engng Gas Turbines Pwr, Trans. ASME*, **116**, 395-401.

(6) **Glahn, A., Kurreck, M., Willmann, M.,** and **Wittig, S.** (1995a) Feasibility study on oil droplet flow investigations inside aero engine bearing chambers – PDPA techniques in combination with numerical approaches, *ASME Paper* (Conference code 43400), p. 9.

(7) **Glahn, A.** and **Wittig, S.** (1995b) Two-phase air/oil flow in aero engine bearing chambers – characterization of oil film flows, *ASME Paper* (Conference code 43400), p. 8.

(8) **Glahn, A., Busam, S.,** and **Wittig, S.** (1997) Local and mean heat transfer coefficients along the internal housing walls of aero engine bearing chambers, *ASME Paper* (Conference code 47341), p. 9.

(9) **Chew, J. W.** (1996) Analysis of the oil film on the inside surface of an aero-engine bearing chamber housing, *ASME Paper* (Conference code 45083), p. 8.

(10) **Launder, B. E.** and **Spalding, D. B.** (1974) The numerical computation of turbulent flows, *Comp. Meth Appl. Mech. Engng*, **3**, pp.269–289.

15

Cavitation in a Pressure-activated Ball Valve

F G Mendonça

Synopsis

A CFD study is undertaken to evaluate the dynamics of a pressure-activated ball valve, including the balance between the fluid-structure force coupling and the predicted effects of cavitation on the hydrodynamics. The fluid-structure moving-mesh coupling and cavitation model are first evaluated in isolation, and then combined in a fully time-accurate and conservative transient calculation with indeterminate valve motion. No experimental validations of the predictions are reported. This study represents the first step towards building confidence in the practical use of cavitation modelling for complex systems.

15.1 Introduction

Performance degradation and structural damage due to cavitation in components such as valves, injectors, pumps, and propellers running under design conditions, is commonplace. Designing against its occurrence usually requires a detailed knowledge of local flow conditions. Predictive techniques, such as CFD, in general offer a cost-effective and time-efficient way of gaining such insight, and in practice contribute significantly to the design cycle, when the inherent physical models have been validated against the application of interest. Recent studies (**1**), (**2**) applied to flow control valves in non-cavitation situations are good examples of such validation. Cavitation adds another and a more complex dimension to CFD predictive requirements.

In flow control devices, one type of which is the focus of this chapter, attached cavitation affects the discharge by reducing the minimum effective flow area. If the cavitation extends across the entire flow cross-section, compressibility effects become important insofar as the local sound speed can reduce between one and two orders of magnitude lower than the equivalent liquid phase value causing the flow to choke (**3**).

A generally applicable, readily available, robust, validated physical model of cavitation for industrial CFD methodologies has been conspicuous by its absence. Towards this target, a recent implementation of a cavitation model (**4**) in STAR-CD (**5**) for incompressible flows represents a significant step forward. In the present work, an attempt is made to undertake, in a methodical way, the prediction of the flow processes in a pressure-actuated ball valve. Transient indeterminate ball valve motion (its motion is governed by the balance between fluid pressure forces and a spring) is combined with the prediction of cavitation. In Section 15.2, the physical processes are described. Section 15.3 contains further details of the CFD methodology used, and in Section 15.4, results are presented and discussed.

15.2 Pressure-activated ball valve

The ball valve illustrated in Fig. 15.1, shown in its closed position, is used as a safety device to relieve pressure build-up in a common-rail fuel injection system. At a critical pressure, the fluid force on the face of the ball equals the spring back force and begins to lift the valve. Discharge through the opening causes the flow to accelerate, and in combination with viscous losses, results in a local reduction of fluid pressure which, if below the critical vapour pressures, will cause the on-set of cavitation. Of particular interest to the designers are the effects of cavitation on the dynamics of the valve, and its contribution to fluid generated noise and component damage.

Fig. 15.1 Three-dimensional image of the valve geometry and close-ups showing the valve in the closed position and mesh structure at 0.1 mm lift

15.3 CFD methodology

Since the inherent nature of cavitation is transient, all the analyses in STAR-CD were run as transient calculations. For stability, the time-step size is automatically adjusted based on the minimum convection time in the flow field. The flow was assumed to be incompressible, turbulent (using the standard k-ε model (**6**)), and isothermal, with no-slip conditions applied at surfaces. First-order practices were used for spatial and temporal discretization. Ideally, second-order practices should be used, but this choice represents a reasonable compromise between the need to integrate CFD into the designer's existing CAD environment and CFD best practices (see Section 15.3.2 below). In the next sub-sections, the cavitation model, mesh and valve motion, and cases studies are described.

15.3.1 Cavitation modelling

The particular method used (**4**) is generally referred to as the 'bubble two phase' (BTF) model. It assumes that cavitation is a result of unstable growth of a spectrum of bubble sizes, R_0 formed through nucleation. Growth occurs when the local pressure falls below the cavitation critical pressure, p_{crit}, defined as

$$p_{crit} = p_v - \frac{4\sigma}{3R} \tag{15.1}$$

where σ is the liquid surface tension coefficient, and R is the actual bubble size. The bubble pressure is determined by an isentropic relationship between the local pressure and p_{crit}, and its size is calculated from bubble radial motion described by Rayleigh's equation (**4**).

The net bubble transport is solved using a volume-of-fluid (VOF) method (**7**) and a higher-order discretization practice (**8**) is used to capture sharp interfaces between the liquid and bubble phases.

15.3.2 Geometry and mesh

The quarter geometry illustrated in Fig. 15.1 was meshed with tetrahedral cells using the designer's existing meshing capability, ANSYS. The resulting mesh is not of optimum quality for CFD calculations, especially in the region where the ball meets the seat (see Fig. 15.1 inset), necessitating the need for first-order discretization practices.

Starting from the meshed geometry with the valve in its closed position, the pre-processor PROSTAR was used to break model at the point of contact between the ball and seat. Mesh layers were extruded from the exposed ball and seat faces to represent the expanding and contracting volume in the valve gap. The final mesh contains 81 597 cells, with 1672 cells on one plane of symmetry, from which an axisymmetric sector was extruded to perform two-dimensional axisymmetric computations. Figure 15.1, (inset) shows the valve with 0.1 mm opening.

15.3.3 Valve and fluid coupled motion

The mechanics of the ball valve system are described by:

Ball diameter	= 3mm
Mass of ball and housing	= 3g
Spring force	= $F_0 + F_1$ * lift

where F_0 is the spring force when the valve is closed, and F_l is the differential force per unit length.

At each time step, the integrated fluid-force on the ball and housing is calculated and balanced against the spring force to determine the net force. Newton's second law is used to compute the ball and housing differential velocity and displacement. The new velocity is applied as a moving wall boundary condition to the surfaces of the ball and valve housing. The new valve position is set by displacing the mesh in the extruded volume between the ball and seat proportionately. STAR-CD then automatically conserves fluxes in the changing volume mesh.

15.3.4 Case studies

First, an equivalence is drawn between three-dimensionality and two-dimensional axisymmetry. Most of the flow features around the region of interest are expected to be axisymmetric. A test scenario was constructed to substantiate this expectation in which the spring force and system inertia were modified to exaggerated valve lift and to allow the system to respond quickly to changes in net force. This was followed by three case studies as summarized below.

15.3.4.1 Mass-flow driven system, indeterminate valve motion, without cavitation to study the mutual coupling between the valve motion and flow processes

A sinusoidally varying inlet mass flow is prescribed at the domain inlet, initially at the mean value of 120 litre/hr with an amplitude of 60 litre/hr. Ambient pressure is set at the outlet, and the transients calculation is run for one full cycle lasting 60 ms. The valve is initially held open at 0.01 mm and it velocity set to zero. The flow field is initialized to the equivalent steady-state solution at that lift and a flow rate of 120 litre/hr. The initial net force drives the valve to open. The time steps are varied automatically during the calculation to be equivalent to a flow field maximum Courant number, based on convection time, of 50.

15.3.4.2 Pressure driven system, fixed valve position with cavitation to study the dynamics of cavitation prediction

Three valve positions are investigated; 0.1 mm, 0.075 mm, and 0.05 mm. Pressure boundaries are prescribed upstream and downstream, at 500 bar and 1 bar, respectively. The calculations were run for a fixed number of time steps, enough to allow cavitation to initiate and to be established. For stability, the computational time steps are automatically adjusted by STAR-CD, based on the minimum convection time in the domain. In general, this represents a Courant number of less than unity.

15.3.4.3 Pressure driven system, indeterminate valve motion, with cavitation

As with the previous study, pressures upstream and downstream are fixed. The initial valve position is set to 0.1 mm and the flow field to static. Initially, the spring force causes the valve to close.

15.4 Results and discussion

Figure 15.2 demonstrates that there is close similarity between the three-dimensional and two-dimensional axisymmetric calculations. The differences in the three-dimensional and two-dimensional force profiles are attributable to the difference in mesh structure in the circumferential direction. This provides sufficient justification to proceed with axisymmetric analyses, since run times are substantially reduced. All subsequent calculations, reported below, were performed on the axisymmetric model.

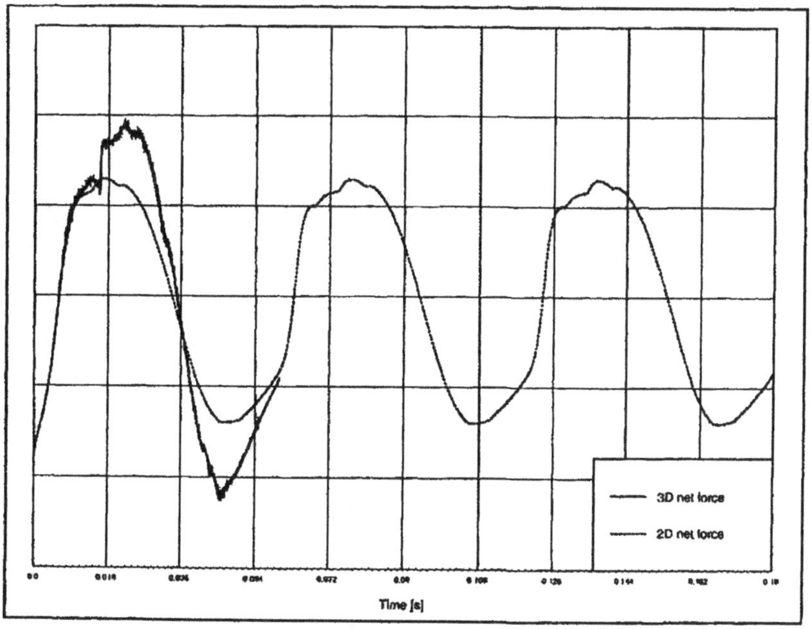

Fig. 15.2 Three-dimensional versus two-dimensional initial study; comparison of lift versus time

15.4.1 Case 1 – mass-flow driven system, indeterminate valve motion driven without cavitation

The dynamic balance of the ball and fluid system are illustrated in Fig. 15.3, showing the variation of ball lift, velocity, net force and upstream flow rate with time. From a lift of 0.01 mm, the initial force causes the valve to open further. The increasing spring back-force then reduces the net lift-force, and its sign reverses, which causes the ball to decelerate, stop and eventually the valve gap begins to close. At this instant, the closure force is at a maximum; it reduces with reducing lift and flow area until, again, the sign of the net force reverses, the ball decelerates to a minimum lift position before the valve begins to open again.

This open-close sequence repeats just over eighteen times through the cycle, corresponding to noise at 300 Hz. The average lift and its amplitude increase and decrease according to the mass flow profile.

It is reasonable to conclude that damping effects due to compressibility are negligible in this simulation since the maximum valve velocity never exceeds 5 per cent of the fluid sound speed. However, the fluid velocities are comparable with the local sound speed.

15.4.2 Case 2 – pressure driven system, fixed valve with cavitation

With an upstream-downstream boundary pressure ratio of 500, Fig. 15.4 illustrates the evolution of fluid pressure force with time for three fixed valve lifts, 0.05 mm, 0.075 mm and 0.1 mm. The behaviour of all three lift-cases are very similar in early part of the simulation, before 0.01 ms. Flow acceleration through the orifice causes a reduction in the fluid pressure force. By this time, one observes the on-set of cavitation and the formation of a vortex downstream of the minimum flow area. Both phenomena increase pressure losses in the system and then reduce the fluid pressure forces which slowly asymptote towards a 'mean-steady' value.

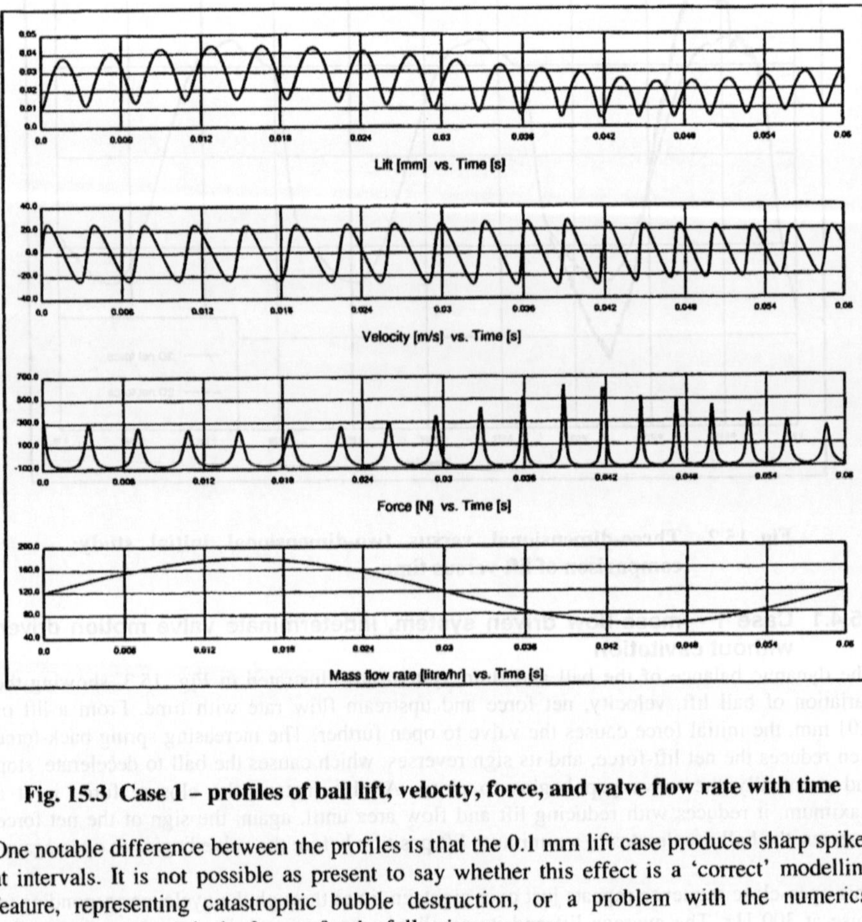

Fig. 15.3 Case 1 – profiles of ball lift, velocity, force, and valve flow rate with time

One notable difference between the profiles is that the 0.1 mm lift case produces sharp spikes at intervals. It is not possible as present to say whether this effect is a 'correct' modelling feature, representing catastrophic bubble destruction, or a problem with the numerics. Indications are that it is the former for the following reasons:

- the spikes do appear at the lower lifts;
- the cavitation patterns are fundamentally different compared with the lower lift cases.

At 0.075 mm and 0.05 mm, the cavitation is not attached to the ball, see Fig. 15.5 (top and middle) which shows the cavitation volume fraction and velocity field development at incremental times of 0.01 ms from the start. Bubble growth initiates in the low pressure orifice region, but the bubbles only increase to a significant volume fraction when they have travelled further downstream. Part of the bubble breaks off and convects with the vortex as it moves downstream. This yields multiple cavitation structures which prevail as the simulations progress.

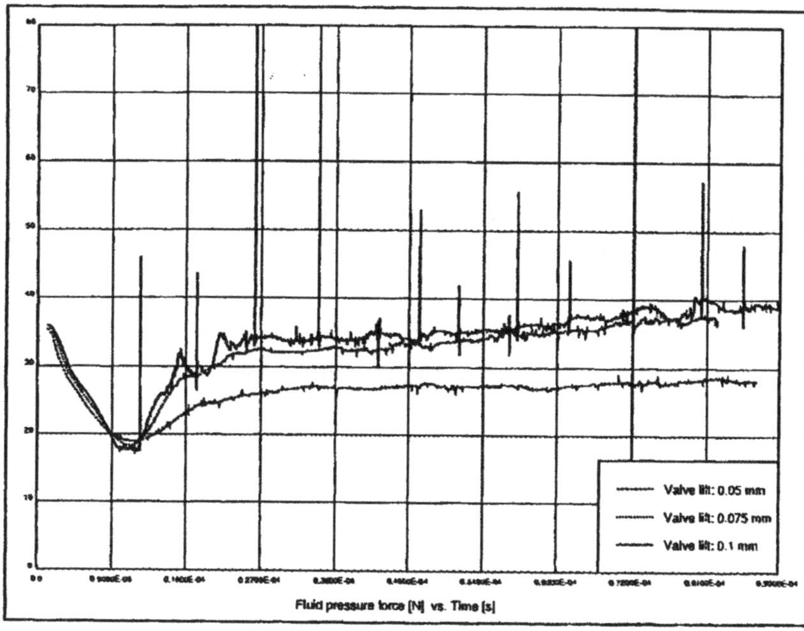

Fig. 15.4 Fluid pressure force history for three steady valve lifts; 0.5 mm, 0.075 mm, and 0.1 mm

In contrast, at 0.1 mm lift depicted in Fig. 15.5 (bottom), one first observes attached cavitation at the upstream edge of the seat which remains throughout the simulation. The developing vortex entrains a cavitation cloud which impacts onto the ball, where bubble clusters continue to convect downstream, detach, or collapse. These unsteady effects on the ball surface could be a cause of the pressure force spikes. Many experimentalists, reported in (**4**), have observed strong interaction between separation induced large-scale vortical structures and cavitation. The low pressure region at the vortex centre provide an environment in which the bubbles remain and grow. Such observations are borne-out in these simulations.

By comparison with Case 1, the time step increments needed to resolve the cavitation dynamics are between one and two orders of magnitude smaller than that required for the fluid dynamics alone.

15.4.3 Case 3 – pressure driven system, indeterminate valve motion with cavitation

The coupling between the valve dynamic movement and cavitation modelling is demonstrated in a system with 100 bar pressure difference, with the valve placed open initially at 0.1 mm and the field at rest; the net force acts to close the valve.

As stated earlier, the time step requirements for cavitation modelling are strict. Typical time step sizes are in the order of 0.01 m. The simulation was run until the lift reduced to 0.05 mm, corresponding to approximately 6 ms duration. Figure 15.6 illustrates the lift, ball velocity, and force histories.

As was seen in the fixed lift study, force spikes can be observed at lifts close to 0.1 mm, but not as the lift reduces when the force profile becomes less noisy.

Figure 15.7 shows a sequence of cavitation volume fraction and flow velocities at lift intervals of 0.0125 mm as the valve closes. The corresponding time interval is too long to capture the development of cavitation clouds, but the location of cavitation relative to the ball and seat can clearly be seen.

By comparing the predictions from case studies 2 and 3, it is reasonable to surmise that the time scales for bubble and cavitation cloud dynamics are much shorter that those for the valve movement. Therefore, fixed lift studies should be sufficient to give an insight into the effect of cavitation for this ball valve system.

15.4.4 Future work

It is intended that the present work will be developed further as follows:

- Acquisition of experimental data to validate the pressure driven simulations described in case studies 2 and 3;
- Use of higher order discretization techniques;
- Modification of the mesh motion to allow removal of layers in the gap between the valve and seat, which will enable indeterminate valve motion and full closure;
- Introduction of compressibility effects;
- Optimization of the temporal algorithm to allow larger time steps when using the cavitation model.

15.5 Conclusions

These analyses demonstrate that CFD can be used as a practical design tool for predicting flows affected by cavitation. Fluid-structure coupled moving mesh methodologies give a useful insight into the valve dynamics, and inferences can be made on noise emissions. Cavitation is shown to be strongly transient, in which cavitation clouds form, detach, entrain into vortical flow structures and collapse.

Transient calculations at fixed lift positions may be sufficient to derive an understanding of the cavitation flow structures. The cavitation time scales have been shown to be much smaller than the valve dynamics.

The CFD predictions have proven to be robust. However, it is recommended that compressibility effects should be included, and also that experimental validations are performed before these methodologies, in isolation and in combination, can be used with confidence.

15.6 Acknowledgements

The author gratefully acknowledges the contributions from his colleagues, Dr. Jianbo Huang and Dr Radenko Drakulic, in bringing this work to publication, and Julien Mazallon, Eaton S A M who provided the geometry and mesh.

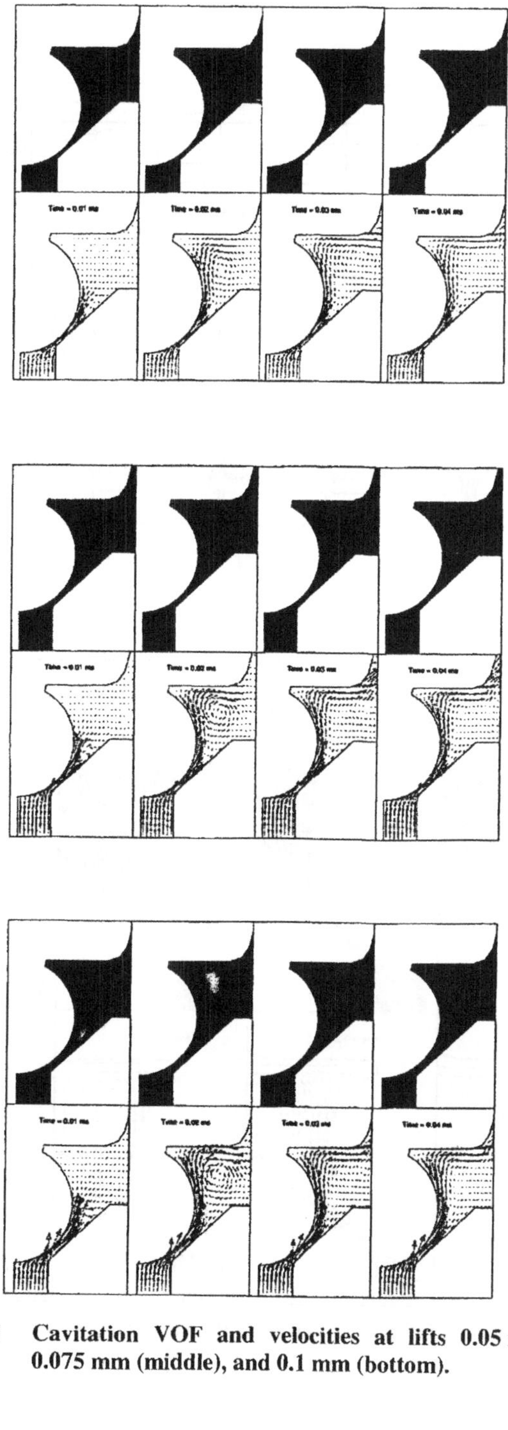

Fig 15.5 Cavitation VOF and velocities at lifts 0.05 mm (top), 0.075 mm (middle), and 0.1 mm (bottom).

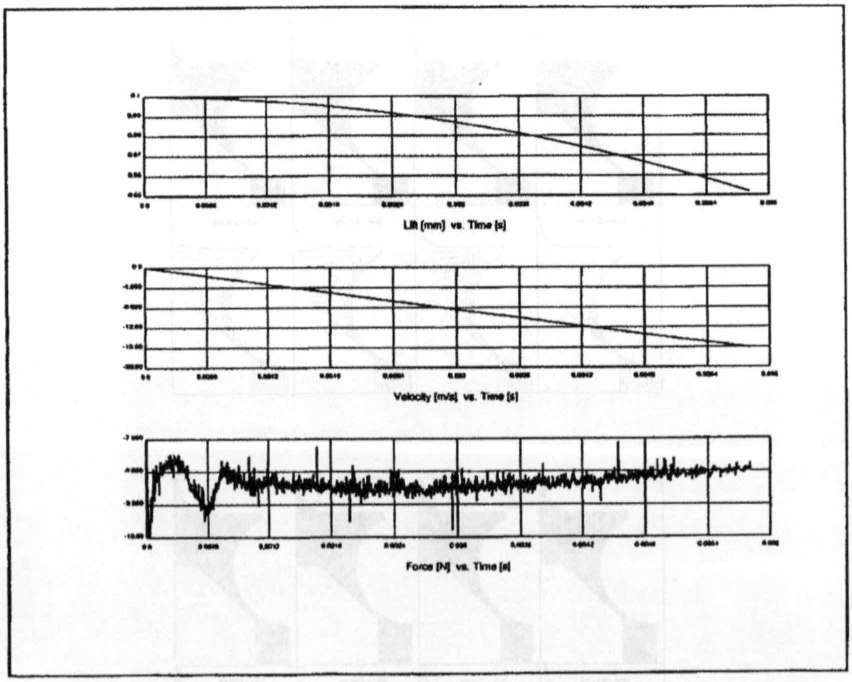

Fig. 15.6 Indeterminate valve motion; valve lift, velocity, and force histories

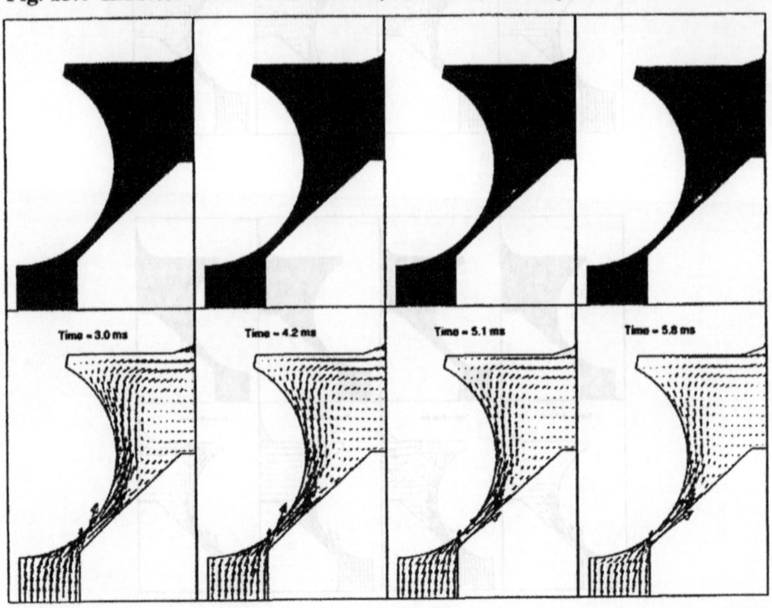

Fig. 15.7 Cavitation VOF and velocities as the valve closes from 0.1 mm to 0.05 mm

15.7 References

(1) **Iudicello, F.** and **Baseley, S.** (1999) CFD modelling of the flow control valve in a hydraulic valve, Bath Workshop on *Power Transmission and Motion Control* (PTMC99).

(2) **Sorensen, H. L.** (1999) Numerical and experimental analysis of flow and fluid force characteristics for hydraulic seat valves with difference in shape, Bath Workshop on *Power Transmission and Motion Control* (PTMC99).

(3) **Halworth, D. C., Maguire, J. M., Rhein, R.,** and **El Tahry, S. H.** (1996) Dynamic fluid flow analysis of oil pumps, SAE/960422.

(4) **Kubota, A., Kato H.,** and **Yamaguchi, H.** (1992) A new modelling of cavitating flows: a numerical study of unsteady cavitation on a hydrofoil section, *J. Fluid Mech.*, **240**, 59–96.

(5) STAR-CD Version 3.10 Methodology and User Guide 1999, Computational Dynamics Limited, London.

(6) **Launder, B.** and **Spalding, D. B.** The numerical computation of turbulent flows, *Comp. Meth. Appl. Mech. Engng*, **3**, p. 269.

(7) **Hirt, C. W.** and **Nichols, B. D.** (1981) Volume of fluid (VOF) method for the dynamics of free boundaries, *J. Comp. Mech.*, **39**, 201–225.

(8) **Ubbink, O.** (1996) *Numerical Prediction of Two Fluid Systems with Sharp Interfaces*, PhD thesis, Imperial College of Science, Technology, and Medicine.

Authors' Index

Subject Index